Testing iOS Apps with HadoopUnit

Rapid Distributed GUI Testing

Synthesis Lectures on Software Engineering

The Synthesis Lectures on Software Engineering publishes 75-150 page publications on all aspects of software design, engineering, and process management.

Testing iOS Apps with HadoopUnit: Rapid Distributed GUI Testing
Scott Tilley and Krissada Dechokul
December 2014

Hard Problems in Software Testing: Solutions Using Testing as a Service (TaaS)
Scott Tilley and Brianna Floss
August 2014

Model-Driven Software Engineering in Practice
Marco Brambilla, Jordi Cabot, Manuel Wimmer
September 2012

Testing iOS Apps with HadoopUnit: Rapid Distributed GUI Testing
Scott Tilley and Krissada Dechokul

ISBN: 978-3-031-01420-8 print
ISBN: 978-3-031-02548-8 ebook

DOI 10.1007/978-3-031-02548-8

A Publication in the Springer series
SYNTHESIS LECTURES ON SOFTWARE ENGINEERING #3

Series ISSN   2328-3319 Print   2328-3327 Electronic

Testing iOS Apps with HadoopUnit

Rapid Distributed GUI Testing

Scott Tilley
Florida Institute of Technology

Krissada Dechokul
Suwat Dechokul Part., Ltd.

SYNTHESIS LECTURES SOFTWARE ENGINEERING #3

ABSTRACT

Smartphone users have come to expect high-quality apps. This has increased the importance of software testing in mobile software development. Unfortunately, testing apps—particularly the GUI—can be very time-consuming. Exercising every user interface element and verifying transitions between different views of the app under test quickly becomes problematic. For example, execution of iOS GUI test suites using Apple's UI Automation framework can take an hour or more if the app's interface is complicated. The longer it takes to run a test, the less frequently the test can be run, which in turn reduces software quality.

This book describes how to accelerate the testing process for iOS apps using HadoopUnit, a distributed test execution environment that leverages the parallelism inherent in the Hadoop platform. HadoopUnit was previously used to run unit and system tests in the cloud. It has been modified to perform GUI testing of iOS apps on a small-scale cluster—a modest computing infrastructure available to almost every developer.

Experimental results have shown that distributed test execution with HadoopUnit can significantly outperform the test execution on a single machine, even if the size of the cluster used for the execution is as small as two nodes. This means that the approach described in this book could be adopted without a huge investment in IT resources. HadoopUnit is a cost-effective solution for reducing lengthy test execution times of system-level GUI testing of iOS apps.

KEYWORDS

software testing, iOS, apps, Hadoop, HadoopUnit, cloud computing, cluster

Contents

Foreword xv

Preface xvii

Acknowledgments xix

Dedication xxi

1 **Introduction** 1
    1.1   GUI Testing of iOS Apps 2
    1.2   Rapid Testing with HadoopUnit 3
    1.3   Related Work 4
        1.3.1   GUI Testing Tools 4
        1.3.2   Distributed Testing Platforms 6

2 **Background** 9
    2.1   Software Testing 9
        2.1.1   Regression Testing 11
        2.1.2   GUI Testing 12
    2.2   UI Automation 12
        2.2.1   UI Automation Script 13
        2.2.2   Command-Line Workflow with UI Automation 15
        2.2.3   Rake 17
        2.2.4   Virtualization 19
    2.3   Hadoop and HadoopUnit 20
        2.3.1   Hadoop 20
        2.3.2   HadoopUnit 23

3 **Using UI Automation with HadoopUnit** 27
    3.1   UI Automation Test Suites 27
        3.1.1   Test Case Design 28
        3.1.2   Test Case Analysis 28
    3.2   HadoopUnit Customization 29
        3.2.1   Operational Environment 29
        3.2.2   Test Results 30

	3.2.3	Revised Architecture	31
3.3		Using HadoopUnit	32
	3.3.1	Test Case List	33
	3.3.2	Rake	34
	3.3.3	Test Execution	35
4		**Rapid GUI Testing of iOS Apps**	**39**
4.1		Experiments	39
	4.1.1	Experiment I	42
	4.1.2	Experiment II	44
	4.1.3	Experiment III	46
4.2		Discussion of Results	48
4.3		Threats to Validity	53
	4.3.1	Test Suites	53
	4.3.2	Hadoop Optimization	53
	4.3.3	Network Issues	54
5		**Summary**	**55**
5.1		Summary of Results	55
	5.1.1	Research Objectives	56
	5.1.2	Research Contributions	56
5.2		Future Work	57
5.3		Concluding Remarks	57

Appendix A
Setting up a HadoopUnit Cluster on Mac OS X 59

Appendix B
HadoopUnit Source Code for iOS GUI Testing 69

References 77

About the Authors 83

Figures

Figure 1.1: App store downloads. 1
Figure 1.2: Overall architecture of third-party testing tools. 5
Figure 2.2: Levels of Testing. 10
Figure 2.3: Iterative development model. 11
Figure 2.4: Sample UI automation test case. 14
Figure 2.5: Sample Xcodebuild command. 15
Figure 2.6: Sample of an instruments command. 16
Figure 2.7: Sample of a Rake task for executing an instruments test. 18
Figure 2.8: Sample of a command to invoke a defined Rake task. 18
Figure 2.9: Sample of a command to invoke a Rake task with--rakefile. 18
Figure 2.10: HDFS architecture. 21
Figure 2.11: Overview of how MapReduce works. 22
Figure 2.12: Overall architecture of HadoopUnit. 24
Figure 3.1: Architecture of HadoopUnit for GUI testing of iOS applications. 32
Figure 3.2: Sample of a test case list. 33
Figure 3.3: Sample of command to transfer a file to the HDFS. 35
Figure 3.4: The Hadoop command to initiate test execution with HadoopUnit. 36
Figure 3.5: The Hadoop command to download files from the HDFS. 37
Figure 4.1: A sample screenshot of the system under test. 40
Figure 4.2: Sample of a test case. 41
Figure 4.3: Code for executing test cases sequentially with a Rake task. 42
Figure 4.4: Sequential execution time on a single machine. 43
Figure 4.5: Concurrent execution time on a 2-nodes cluster. 45
Figure 4.6: Concurrent execution time on a 4-node cluster. 47
Figure 4.7: Execution time comparison of the three experiments. 49
Figure 4.8: Total test execution time approximation equation. 50
Figure 4.9: Ideal case for test execution time approximation equation. 50

Tables

Table 4.1: Sequential execution time on a single machine (in seconds)   43

Table 4.2: Concurrent execution time on a 2-node cluster (in seconds)   44

Table 4.3: Concurrent execution time on a 4-node cluster (in seconds)   46

Table 4.4: Performance comparisons of the three experiments (in seconds)   48

Table 4.5: Performance factors over sequential execution   49

Foreword

Software quality has never been as important as it is today. A tidal wave of new software development is seen in mobile applications, which are quickly becoming ubiquitous. They are used for many purposes, from enjoyable entertainment to safety critical applications used by first responders.

At the same time, the bar to entering the mobile application marketplace is very low. Anyone with the desire to create a mobile application can do so with freely available tools in a rather short amount of time. A quick browse through the Apple App Store reveals that many applications fall short of user expectations: there are many apps in the store with three stars or fewer.

Creating a mobile application is one thing. Testing it to ensure that it works correctly, is secure, and achieves high usability factors is another thing altogether.

First, there is the plethora of mobile device platforms and appliances—each of which has its own configuration and behavior differences.

Second, the time required to test a moderately featured application just one time on one platform is substantial. Manual testing can be tedious and automated testing simulating user actions can relieve some of the tedium, but it still requires time to design and perform functional tests.

Third, mobile applications undergo constant change. The constant march of change fixes bugs and introduces new features—and new defects. Therefore, testing should include not only testing bug fixes and new features, but also regression testing of the unchanged features. However, this all takes precious time.

Finally, consider that many mobile application developers consist of small teams, perhaps just an individual, with limited funds and resources for testing. So, the customers get to be the "testers," except they use the applications for important real-world tasks like arranging travel, managing finances, tracking severe weather, and navigating roads. The customers don't consider themselves as testers. They are customers and even though they might not have paid any money for an app, they don't like to waste their time on defective ones. Also, when mobile applications fail during important tasks, defects are more than a mere inconvenience—they impact lives in a negative way.

While developers may perform bug fixes, customers often abandon apps quickly due to a bad experience. It is very easy to delete an app from an iOS device, so developers who wish to have a successful app in the App Store need to understand the detrimental impact of defects on their success.

That's why this book is important for mobile application developers and testers. HadoopUnit offers a solution for testing mobile apps on the iOS platform that is not only free, but reduces test time dramatically. There will still be a need for usability and compatibility testing, but the more troubling defects are those that impact mobile application reliability.

Scott Tilley and Krissada Dechokul have done an excellent job with this book in describing in detail how to design and implement tests in HadoopUnit. My hope is that everyone involved in developing iOS applications will read this book and create more reliable and robust applications that get five-star reviews.

Randall W. Rice
CSTE, CSQA, CTAL
Founder, Principal Consultant, and Vice-President of Research and Development
Rice Consulting Services, Inc.

Preface

The app ecosystem is enormous. We have grown dependent on smartphone apps for almost every aspect of our lives. When the apps don't perform as expected, the consequences can range from midly irritating to life-threatening. For this reason, testing smartphone apps is very important—particularly the GUI that is the main window into the app's functionality. Unfortuanately, GUI testing can be a time-consuming process, which leads to fewer tests being run, further exacerbating the app's quality problems.

This book focuses on the specific problem of GUI testing for iOS apps found on Apple's products such as the iPhone and the iPad. Apple provides developers with a testing framework called UI Automation, but its capabilities are limited in terms of speeding up the testing process. The result is that GUI testing of complex iOS apps can take many hours.

The solution proposed here is to leverage the parallelism inherent in the Hadoop distributed platform to provide an environment for concurrent test execution. The approach builds upon an existing system, called HadoopUnit, that was previously used to reduce regression testing time of large JUnit test suites. HadoopUnit has been customized to drive UI Automation test cases in a manner that is easy for developers and testers to adopt, yet provides measurable improvement in test case execition times.

WHAT IS UNIQUE ABOUT THIS BOOK?

This book represents the continuation of research that began in 2009 on addressing the problem of execution times for large regression test suites. The first of this work were the HadoopUnit distributed execition environment and the SMART-T migration decision framework [60]. The work also led to the creation of a new community of researchers and practicioners interesting in software testing in the cloud (STITC) [59][62].

The STITC project evolved to examine the applicability of testing as a service (TaaS) to hard problems in software testing (HPST) [61][58]. TaaS is a promising new development that offers a service-oriented interface to the testing capabilities provided by an environement like HadoopUnit. The number one problem found in the HPST project was education & training, and it's currently an open question where TaaS may help alleviate this timeless challenge.

GUI testing of iOS apps is a timely update and specific instance of the classic regression testing problem, and one that HadoopUnit is well suited to address. The research reported in this book is unique is its application of an advanced environment such as HadoopUnit for concurrent testing in manner that is accessible to almost all developers and testers—even if they are of modest means. A simple two-node cluster is all that is needed to realize significant testing benefits.

WHO SHOULD READ THIS BOOK?

Anyone who is involved in GUI testing of iOS apps will find the material presented in this book valuable. This is particularly true for testers, but developers, managers, and even end-users can benefit from understanding the challenges faced when using the UI Automation framework and the possible benefits of using the customized HadoopUnit to address these challenges.

Modern software engineering—and app development in particular—involves the use of sophisticated IDEs and integrated coding platforms. Many of these tools are moving to the cloud. An understanding of how Hadoop can be used in the domain of GUI testing provides valuable insight into the power of the MapReduce programmign paradigm.

OUTLINE OF THE BOOK

Chapter 1 discusses the challenges of GUI testing with UI Automation, outlines prior results for rapid testing using HadoopUnit, and outlines related work in the areas of GUI testing tools and distributed testing platforms. Chapter 2 provides background information on the challenges of software testing in general, and GUI testing of iOS apps with UI Automation in particular, and summarizes the Hadoop platform and the HadoopUnit distributed test execition environment. Chapter 3 outlines how HadoopUnit can be used to drive the UI Automation framework to faciliate parallel test execution. Chapter 4 details three experiments in rapid GUI testing of iOS apps using the customized HadoopUnit. Lastly, Chapter 5 summarizes the main results, objectives, and contributions of this work and outlines possile avenues of further investiation.

The book also contains two appendices. Appendix A describes how to set up a HadoopUnit cluster on Mac OS X. Appendix B provides source code samples for HadoopUnit, suitably modified for iOS GUI testing with UI Automation.

Scott Tilley
Melbourne, FL

Krissada Dechokul
Bangkok, Thailand

October 2014

Acknowledgments

We are indebted to everyone who helped develop HadoopUnit—the platform upon which this research is built: Tauhida Parveen, Eric Bower, colleagues at Yahoo!, collaborators at SAP, and members of the global STITC community.

Our thanks to Apple for making the excellent UI Automation framework available for free as part of the Xcode development environment.

We appreciate the the invaluable comments privided by the book's reviewers. Their suggestions helped improve the text. Any remaining errors or omissions are ours alone.

We are grateful to the Florida Institute of Technology for supporting this research.

Finally, our gratitude to Morgan & Claypool for their guidance and patience in helping us publish the results of our work.

Dedication

To Miel
— Scott Tilley

To my parents
— Krissada Dechokul

CHAPTER 1

Introduction

Mobile applications have become an important part of the software development industry and a driver of the overall technology economy. In June 2014, Apple announced that 75 billion apps had been downloaded from its App Store [53]. As shown in Figure 1.1, Apple is on track for 85 billion downloads by the end of October. Gartner predicts the total number of mobile apps downloaded across all platforms to double by 2016 [22].

Cumulative number of apps downloaded from the Apple App Store from June 2008 to October 2014 (in billions)

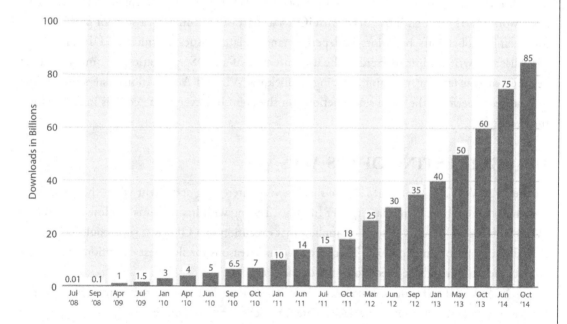

Figure 1.1: App store downloads [53].

One of the main factors behind these impressive numbers is the growing use of smartphones and tablets. Smartphones will soon pass personal computers to become the dominant computing platform worldwide [49]. With so many apps in use, quality becomes an important issue. A game crashing is one thing; an app interacting with essential systems is something else altogether.

In July 2014, Apple and IBM entered into a strategic partnership to develop business-focused apps for the enterprise [45]. In October 2014, IBM and SAP entered into a strategic partnership wherein IBM will provide the cloud infrastructure for SAP's enterprise applications, including its growing number of mobile apps [24]. These agreements signal a clear shift in the app ecosystem, away from a purely consumer-oriented world to one that embraces corporate users as well. This shift will only increase the need for high-quality mobile apps.

One way that helps ensure quality of mobile applications is through testing. Software testing is one of the main activities in the software engineering life cycle (requirements, design, construction, testing, and maintenance). Testing is a very important process that helps assess the functionality and ensure the quality of software before being released by identifying defects and problems [8]. System-level GUI testing, which is a primary focus of this work, is especially important in mobile application testing because it tests an application after all components has been integrated to ensure that the application behaves as intended [47].

On the iOS platform, GUI testing can be done through the use of Apple's UI Automation framework [4]. It was first introduced with iOS 4 as part of Instruments, a suite of profiling tools for tracing application's behaviors and performance related issues at runtime. UI Automation allow users to write scripts to exercise the user interface of an iOS application by simulating users' gestures, such as tapping, pinching, swiping, or flicking. With UI Automation, testers do not have to repeatedly perform the same set of actions on the app GUI every time code is integrated into the baseline.

1.1   GUI TESTING OF IOS APPS

At the 2010 WWDC conference Apple made a very interesting statement when they first introduced UI Automation: "Automating User Interface Testing with Instruments or How to Find Bugs While You Sleep" [2]. This statement implied that execution of a GUI test suite could take a considerable amount of time to finish. There are two main reasons for slow test execution: the number of test cases to run and the execution time of each test case.

A good GUI test suite covers all functionality of an application accessible through its user interface and exercises every user interface element presented to users on screen. The user interface of a mobile application as well as the variations through which users can interact with the application can grow large in number and the amount of test cases to cover in a test suite increases accordingly [13]. Moreover, the test suite continues to grow over the course of software development process, especially when agile methods are used: more test cases are added each iteration. The testing also needs to account for different versions of operating systems, device orientations, and form factors of mobile devices.

The execution speed of GUI testing also can be very slow and time-consuming. This is because testing through the GUI proceeds in real time, at end-user speeds, not processor speeds [39]. This type of testing asserts the behaviors of an application from the outside, just as the users would actually see. This means that the testing tool needs to constantly wait for the rendering of user interface elements and animations, or wait for network delays throughout the test execution. Combining the slow test execution speed with the total number of test cases results in the test process being tediously slow.

GUI test execution of a realistic test suite can take an hour or more to execute, and if the user interface of the application is complicated, the time required for a test execution to complete might go beyond the time and resources available for a project. This limits how many times testing can be performed, which in turn affects app quality.

To solve the problems presented by slow-running large test suites is more complicated than simply having multiple separated processes executing more test scripts on a single machine. One reason is that with UI Automation, only one test script could be run at a time. Moreover, only one instance of iOS simulator could be run at a time [4]. A different solution is needed.

1.2   RAPID TESTING WITH HadoopUnit

There are many solutions proposed to solve the test execution time problem. One way is to reduce the number of test cases to run. There are many researchers [36, 23] who propose various approaches to reduce the number of test cases to run. Instead of executing the whole regression test suite, selective re-test techniques attempt to execute only a subset of test cases in a test suite. There are several such techniques, such as test case selection [51], test case prioritization [66], and test suite reduction [25]. However, these kinds of techniques are complicated and can be computationally expensive, as decisions have to be made on which test cases not to run. With such decisions, some of the regression bugs might have been missed with the dropped test cases if the analysis is not flawless. The re-test all approach is more straightforward in that it avoids such errors [23]. Further discussion on regression test selection is provided in Section 2.1.1.

Another approach to mitigate the problem is to move to a faster test execution environment. A pre-existing framework, called HadoopUnit [46], was designed to solve this problem. In previous studies, HadoopUnit provided an infrastructure for executing JUnit test cases concurrently in a cluster or in the cloud to speed up the test execution process. The result from the previous studies have shown that using a medium-scale cluster of 150 nodes, the test execution runtime of Hadoop's JUnit test suite with 230 test cases could be impressively improved for up to approximately 55 times faster [60] by executing a single test per node. However, because not everyone has access to large amount of IT resources, even with a medium cluster as shown from the previous studies, there is also a study of HadoopUnit on a smaller-scale of cluster of 4 nodes and the result have shown that such impressive performance improvement is not always guaranteed. Some modifications had been made to HadoopUnit to further fine-tune its performance by letting each node to execute more than one test concurrently [9].

The structure of the HadoopUnit platform and the problems that it solved by reducing the execution time of JUnit test suite make it a suitable candidate as a platform for system-level GUI testing of iOS applications. In order to migrate testing to a distributed execution environment, several artifacts involved in the testing process of the current test execution environment must be identified. The important artifacts in system-level GUI testing of iOS applications include a set of test cases that composes a test suite, the UI Automation testing tool, a trace document capturing the test results of each test execution, and the operational test environment. These artifacts are studied in this book for feasibility in migrating to a new environment, with HadoopUnit customized accordingly as needed.

1.3   RELATED WORK

There are several other GUI testing tools and distributed testing platforms related to the work presented in this book. The GUI testing tools described below offer alternatives to Apple's UI Automation, but come with their own caveats. There are fewer choices when it comes to distributed testing platforms currently available, but that may change if Testing as a Service becomes more prevalent.

1.3.1   GUI TESTING TOOLS

The primary reason that UI Automation was selected as a testing tool for GUI testing tool in this book was because it works without requiring software engineers to install any additional software or customize their Xcode projects. However, there are several third-party testing tools available that could be used GUI testing as well. The main reason to consider a third-party testing tool is when the team was already familiar with the tools or language used by the tools. Although they rely on different technologies and implementations, they share similar concepts on how these testing tools work together.

As shown in Figure 1.2, there are three common components in these testing tools: an agent, a script or feature, and an IDE.

1. An Agent: A framework or a small server that will be installed to run inside the system under test. The installation needs to be performed on an iOS project manually through the Xcode IDE.

2. A Script or a Feature: The main set of test scripts for the system under test. Programming languages used for each tool will be different based on the underlying technologies

used by the testing tool. For example, Objective-C [5], JavaScript [42], CoffeeScript [6], or Cucumber [26].

3. An IDE: This component is separate piece of software running on a tester machine or remote server that helps record the interactions, execute the scripts, and issue commands that interact with the system under test through the agent component installed inside the application.

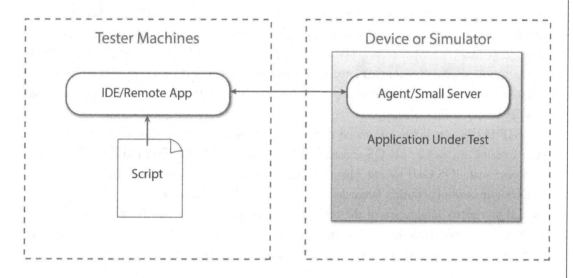

Figure 1.2: Overall architecture of third-party testing tools.

GUITAR [43] is a framework for performing automated GUI testing. The GUITAR framework has several notable features including automated test case generation and execution. It consists of four main components: (1) Ripper for generating a structure model of the GUI of the system under test; (2) Graph Converter for converting the structure model into a graph; (3) the Test Case Generator for generating test cases based on the graph; and (4) Replayer for automated executing the generated test cases. GUITAR has a plugin-based architecture meaning that a specific plugins are needed to support different platforms. GUITAR supports multiple platforms, such as Java, Web, and Android. iPhone GUITAR [19] is still under-development to support GUI testing of iOS application. One weakness of GUITAR is that it does not support any manual test cases development (both scripted and captured) and leads instead towards automating the whole process.

MonkeyTalk [11] is a free, open-source, object-based recording tool from Gollila Logic that examines the app's code to understand the underlying interactions. This approach is supposed to be more accurate that image-based recording that detects pixel on the screen. It also results in a more readable and maintainable script when compared to scripts that rely on image recognition technology. MonkeyTalk has its own IDE to construct or record a new test script and playback existing scripts. There are a lot of actions and gestures to select from, including wait time specified in milliseconds. MonkeyTalk uses its own proprietary MonkeyTalk Scripts for constructing test cases while still supporting JavaScript. MonkeyTalk also support command-line test execution through either Ant Runner or Java Runner.

Frank [57] is a tool for automated iOS testing by Pete Hodgson. It is described by the tool's developers as "Selenium for native iOS apps." Frank uses a combination of Cucumber and JSON commands that are sent to a server running inside the native application. The benefit of using natural language like Cucumber is its readability for non-technical stakeholders in the project. Frank includes a tool called Symbiote that let users explore the system under test via a Web browser. Its focus, however, is specifically on running a simulator, making it hard to run the test suite on the actual device.

KIF [52] is a tool for automated iOS testing created by Square, the company behind the Square Credit Card Reader for iOS and Android devices. Like MonkeyTalk and Frank, KIF allows users to automate iOS GUI testing. However, the tool is based on Objective-C and is built on top of OCUnit, instead of scripting languages like JavaScript, Cucumber, or MonkeyTalk Script. This does lead to a better integration of the test suite; it can access almost anything inside the system under test. However, KIF is more favorable for teams that consist mainly of developers (as opposed to testers) because programming knowledge is required to construct the test cases.

1.3.2   DISTRIBUTED TESTING PLATFORMS

There are several projects that implement a distributed testing platform or framework for software testing and share a similar focus of reducing test execution times for a large regression test suite. There are a few commercial offerings beginning to appear, but they are not suitable for use with UI Automation.

CloudTesting [16] is a framework that provides a solution to parallelize the execution of a test suite over a distributed infrastructure. It has a similar goal to HadoopUnit, which is to improve the performance of the test execution process by distributing the execution of a test suite over a networked infrastructure such as cloud or a private cluster. However, CloudTesting takes a different approach by integrating the framework with development tools used by software developers or testers with the goal to encapsulate all of the underlying work from them. The framework was divided into several different components.

- Configuration: This component takes care of the underlying configuration for the project, such as the paths, hosts in the cluster, and load balancing parameters between them, as well as other settings necessary to set up the test environment.

- Reflection: This component takes care of the test case extraction process that extracts all the test cases to be executed under the distributed infrastructure. The result from this component is used by the next component, Distribution.

- Distribution: This component sits in the middle of GridUnit, between the framework and the distributed infrastructure, and takes care of distributing the execution of test suites and moving test results back and forth between them.

- Connection: This component provides communications with the distributed infrastructure. It is used by the Distribution component to manage each test and its test result back to the client.

- Log: This component records every event that occurred during the test process.

- Main: This component encapsulates other components to make the distributed testing process transparent to the users.

All of these components are packaged as a separated plugin for the IDE used by software developers and testers. The current implementation only supports the Eclipse IDE and the Amazon Web Services infrastructure. Most of the responsibility that these components provide was already available by Hadoop. To support the other development IDE such as Xcode, another plugin for the IDE needs to be separately developed and so could not be used in the context of this work.

GridUnit [17, 18] is a grid-based test execution tool for distributing the execution of JUnit test cases to a computational grid environment. GridUnit was developed on top of OurGrid [44], a peer-to-peer computational grid that provide access to an execution environment to run parallel applications. GridUnit provides functionality similar to HadoopUnit in that it provides automatic test case distribution, test load distribution, and test execution control. However, GridUnit require users to develop a test runner to schedule jobs; with HadoopUnit, the underlying Hadoop platform handles this automatically. Moreover, in its current form, GridUnit supports test case executions of JUnit written in Java only, while HadoopUnit was designed to be able to execute test cases independent of programming language.

CHAPTER 2

Background

This chapter provides background information of three key topics related to this work: software testing, UI Automation, and Hadoop and HadoopUnit.

2.1 SOFTWARE TESTING

Software testing is a very important activity in the software development lifecycle. It involves a technical investigation and evaluation of a software product to find information about its quality [34]. It is one of the main activities in the classic software engineering life cycle like the waterfall model [54] (Figure 2.1).

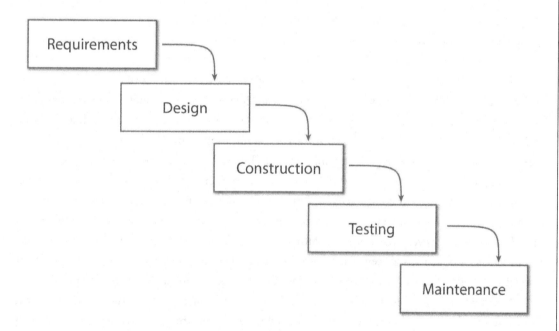

Figure 2.1: Software engineering life cycle—waterfall model.

Software testing may occur at any different level throughout a software development life cycle. According to the ISTQB, there are 4 levels of testing: component testing (or unit testing), integration testing, system testing, and acceptance testing [8] (Figure 2.2).

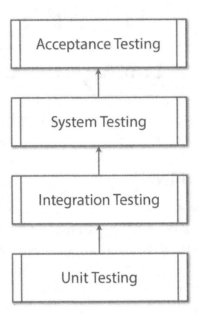

Figure 2.2: Levels of Testing.

Component testing (also known as unit testing) is a level of testing that runs at the lowest level down to individual function or method of a particular software product. There are many different types of tools that help testing at this level, for example, a unit test framework like JUnit for testing Java applications, CppUnit for testing C++ applications, and OCUnit for testing Objective-C applications. Even simple debugging tools can be used for component testing [21].

Integration testing studies how two or more components work together by inspecting the interfaces between different components or between different parts of a system under test. System testing tests the behavior of a system or a software product by considering them as a complete system, after integrating every component together, in an attempt to demonstrate whether or not the system meets its objectives. Acceptance testing is a level of testing that tests a system or a software product with respect to the users' needs to determine whether or not it should be accepted.

Software testing raises the level of confidence about the quality of a software product before it's released to the users or customers. It can also be used to ensure that software works correctly after being migrated to a new operating environment. Alessandro and Filippo illustrated the importance of software testing in this context, where testing can used to find potential divergences between the system under migration and the newly migrated system by performing regression testing [40]. They emphasized the importance of creating a system-level test suite used to test the newly migrated system on every iteration of migration to confirm the correctness of the system and the equivalence with the previous version of the system. System-level testing helps reassure that changes in detailed implementation of the underlying components, such as code refactoring or even internal architectural changes, will not affect the overall functionality of the system.

2.1.1 REGRESSION TESTING

IEEE defines regression testing as a process of rerunning the test of a system to ensure that any modification of the system does not introduce any unintended side effects [28, 37]. It involves executing those test cases that have previously been executed to confirm that any modifications of the software or the environment have not broken any part of the application. Regression testing can be performed at any of the four levels of testing.

Regression testing is generally performed whenever a new version of an application is produced or a new version of environment (e.g., an operating system upgrade) is released. Performing regression testing helps ensure that a new release of an application did not introduce any form of regression errors. For instance, a fix or modification to one component might cause another component to break, or cause a previous bug fix to fail [32].

Regression testing is particularly important in iterative software development, where a software project is broken down into several short phases [8](Figure 2.3). In iterative life cycles, newly introduced functionality will need testing and all existing functionality will need regression testing, meaning more testing is required on each subsequent iteration. This leads to larger regression test suites due to the accumulation of test cases over time. With large test suites, performing regression testing can become costly and it might not always feasible to run everything due to the excessively long execution times incurred.

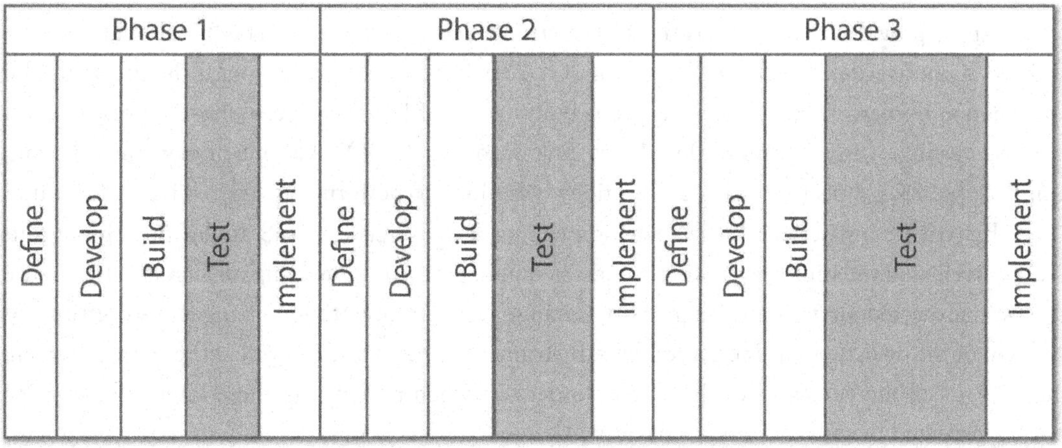

Figure 2.3: Iterative development model.

2.1.2   GUI TESTING

According to Ayman et al., GUI testing is a type of testing that validates the visual properties of GUI elements and the functionalities accessible through them [30]. GUI testing is performed on an application to verify that it functions and behaves correctly, whether or not it does what it was supposed to do, and if it provides good user experience to users [41].

The GUI is a very important part of an application because it acts as a bridge that connects the application and its users. Unlike unit testing, which focuses on the component level (an individual unit inside of an application), GUI testing focuses on the system level—the user interface from the outside of an application, just as the users would see.

GUI testing involves walking through every user interface element of an application and following all workflows through the interface to accomplish a particular task. This means there can be a lot of test cases to cover in order to test every possible variation that users could experience. This type of testing can be time-consuming because every interaction has to pass through the GUI of the application, for the entire test suite, in real (not simulated or accelerated) time [39].

2.2   UI AUTOMATION

Manual GUI testing of an application involves launching and walking through every feature that an application provides through its GUI, one by one, until all possible variations are covered. This can be cumbersome and error-prone if the application is complicated enough. There will be mistakes and some specific paths or variations that users could take would be left untested. Nothing can replace manual testing activities by skilled testers; however, such repetitive tasks can be automated using available automated testing tools [47].

Apple provides the Xcode IDE for free. It is a powerful suite of tools for developing iOS and OS X application. On the iOS platform, GUI testing can be done through the use of the UI Automation framework. UI Automation was built on top of JavaScript to perform automated user interface testing of an iOS application. It was first introduced in iOS 4 as one of many instruments, a suite of profiling tools for tracing application's behavior and performance related issue at runtime.

By writing test scripts, UI Automation can used to exercise an iOS application through its GUI as well as simulating user gestures, such as tapping, pinching, swiping, or flicking. It can also handle unexpected alerts, simulating multi-tasking, handling orientation changes, and setting the location of the device to specific coordinates. Instruments creates a report called the trace document that records all the behaviors of the system under test when performing these interactions, which can be inspected by testers for misbehaviors [4].

2.2.1   UI AUTOMATION SCRIPT

Apple represents all user interface elements of an application through the Accessibility framework. UI Automation communicates with a system under test using assistive technologies such as VoiceOver that reads aloud all the user interface elements and gives visually impaired users a clue on how to navigate an application. Anything that the Accessibility framework can see is accessible by UI Automation [47].

User interface elements displayed on a screen of an iOS application are accessible through the JavaScript API provided by UI Automation. Although the language and the syntax are the same, there are some differences between the JavaScript library used by the UI Automation framework and the JavaScript library used by typical web browsers. For example, there is no concept of document object model (DOM) in the UI Automation framework and there are some additional APIs provided to facilitate creating a test suite, such as the "#import directive" and methods to filter array of elements.

Instruments also provides tools that help recording and play back interaction with an application, while also generating a line of automation script that perform the recorded actions automatically. This feature greatly helps testers building their test cases faster. However, it doesn't record timing of the recorded actions, although delays could be added manually if timing is desired.

Since it is built on top of JavaScript, UI Automation gives testers flexibility in writing test cases to test their applications. The test script can be composed of a collection of statements to assert the behavior of an application under test or it can be grouped together into a simple JavaScript function and get invoked like a normal JavaScript function. An example of a simple test case can be seen in Figure 2.4. The test case asserts that a table cell was actually removed from the table after simulating various actions and gestures.

```
      var target = UIATarget.localTarget();
      var app = target.frontMostApp();
      var window = app.mainWindow();

  function testcase() {
          UIALogger.logStart("Simple test case");

          var navigationBar = app.navigationBar();
          var editButton = navigationBar.leftButton();
          editButton.tap();

          // Grab a table cell with the name "Test"
          var cells = window.tableViews()["InfoList"].cells();
          var testCell = cells["Test"];
          var deleteSwitch = testCell.switches()[0];
          deleteSwitch.tap();
          var deleteButton = testCell.buttons()[0];
          deleteButton.tap();

          if (testCell.isValid()) UIALogger.logFail("Failed");
          else UIALogger.logPass("Passed");
  }
  testcase();
```

Figure 2.4: Sample UI automation test case.

UI Automation provides various means to report whether a test case passed or failed by using simple logging functions, as shown in Figure 2.4. There are two main types of logging functions: logging with Test Status and logging with Severity Levels. Logging with Test Status functions are those functions that indicate that a test has completed successfully, unsuccessfully, or terminated abnormally. Examples of the functions of this type are logStart, logPass, logFail, or logIssue. Logging with Severity Levels function allows the tester to write debugging message with a specific level of severity. Examples of the function include logMessage, logError, logWarning, and logDebug.

Instruments also automatically grabs a screenshot of the current state of the system under test whenever a test case fails, so testers could see what the screen looked like to help determine what went wrong when the test case failed. A function to capture the screenshot can be invoked manually at any point in the test case. UI Automation records log messages and stores these screenshots in a trace document so that the testers can later inspect the document for misbehaviors.

Although there is no built-in support for assertions in UI Automation, UI Automation scripts can be extended by importing an external library like TuneUp JS that provides a collection of assertion functions to be used to provide extra functionality [63]. The written test script can be used to test both iPhone and iPad application on a simulator and also can be used to test on an actual device that connected to Xcode [4].

2.2.2 COMMAND-LINE WORKFLOW WITH UI AUTOMATION

Taken together, the Xcode IDE and Instruments provide an integrated user interface for all development activities from coding to testing. However, in order to automate the testing process, the test has to be executed through the command-line interface. There are two important commands related to GUI testing of iOS applications and are used throughout this work: xcodebuild and instruments.

xcodebuild is a command to build an Xcode project through the command-line terminal. A typical structure of this command is shown in Figure 2.5. (The backslash at the end of each line is just to break the command into multiple lines for the readability.)

```
xcodebuild \
        -project PROJECT_NAME.xcodeproj \
        -scheme PROJECT_SCHEME_NAME \
        -configuration Release \
        -sdk iphonesimulator \
        CONFIGURATION_BUILD_DIR=/tmp/PROJECT_DIRECTORY \
        TARGETED_DEVICE_FAMILY=1 \
        build
```

Figure 2.5: Sample Xcodebuild command.

- The **-project** option is to specify the project file of the Xcode project.

- The **-scheme** option is to specify the scheme for Xcode to build a project. A scheme is a collection of configurations, build setting, and targets that Xcode uses when building or testing an application. There can be multiple schemes in an Xcode project but only one can be active at a time (it can be specified with this option).

- The **-configuration** option instructs the compiler which build configuration to use. Xcode's Profile action builds an app for the Release configuration, so for testing, Release is used.

- The **CONFIGURATION_BUILD_DIR** is an Xcode configuration setting to specify which directory the compiler should put all the build-related files into. This directory will also be used by the instruments command to perform testing.

- The **TARGETED_DEVICE_FAMILY** is another Xcode configuration setting to specify which device family to target. The value of 1 means the targeted device family is iPhone or iPod touch while the value of 2 means the targeted device family is iPad. The value of 1,2 means the app is universal.

- The last **build** option is just to instruct Xcode to build this project.

After the Xcode project has been built with xcodebuild command, the application is ready for testing.

instruments is a command to start the Instruments tool through the command-line terminal. A typical structure of the instruments command is shown in Figure 2.6.

```
instruments \
        -t automation/TEMPLATE.tracetemplate \
        -D automation_results/TRACE \
        /tmp/PROJECT_DIRECTORY/SUT.app \
        -e UIARESULTPATH automation_results \
        -e UIASCRIPT automation/TEST_SUITE.js
```

Figure 2.6: Sample of an instruments command.

- The **-t** option is to tell Instruments which trace document template to use for this test execution. A trace document tells Instruments to start a profiling tool for the task, which, in this case, is the UI Automation tool. In the example, the trace-document template is TEMPLATE.tracetemplate and is stored in the automation directory.

- The **-D** option tells Instruments the directory in which it can store all the trace documents. That directory needs to be created beforehand, otherwise the tool won't start. In the example, the directory is automation_results and the trace document is named TRACE. The trace document is very important as Instruments stores all information about the testing into the file, including all log messages and screenshots of the application under test that will give clues to testers what's wrong with the application when it fails. When there is a problem with test execution, this directory is where testers can get all the test-related information.

- The following line is the app-bundle of the system under test that was previously built using the xcodebuild command. In the example, the app-bundle is named SUT.app and is stored in /tmp/PROJECT_DIRECTORY.

- The **-e UIARESULTSPATH** is to specify an environment variable to tell UI Automation the location to write an XML copy of the automation trace log into.

- The **-e UIASCRIPT** is to specify an environment variable to tell UI Automation the path to the test script to be used for testing.

The basic structure of these commands indicates what information testers need for each test execution and that HadoopUnit needs to track.

2.2.3   RAKE

Rake is a Ruby build tool with capabilities similar to Make [64]. Rake works on Mac OS X without requiring any additional programs to be installed. With Rake, all complexity of building an Xcode project with xcodebuild and running the test with Instruments commands can be hidden in a very simple Rake task. There also are some useful features of Rake that can be added into the process. Those features include:

- creating or removing a directory;

- storing common paths into variables;

- processing or constructing string;

- sequentially executing a command after another;

- conditionally executing commands based on given parameters; and

- killing process after finish executing a task.

The fragment of code shown in Figure 2.7 is a sample of a Rake task defined for executing a test with Instruments. The code starts by defining a number of common paths of an Xcode project. Then the code defines a Rake task named **test,** which accepts a UI Automation script's file name before starts building and performing a test, respectively. After testing is finished, the simulator process is terminated.

```
BUILD_DIR                       = "/tmp/PROJECT_DIRECTORY"
APP_BUNDLE                      = "#{BUILD_DIR}/AUT.app"
AUTOMATION_TEMPLATE             = "automation/TEMPLATE.tracetemplate"
RESULTS_PATH                    = "automation_results"
OUTPUT_TRACE_DOCUMENT           = "#{RESULTS_PATH}/Trace"
PROJECT_DIR                     = "PROJECT_NAME.xcodeproj"
BUILD_SCHEME                    = "PROJECT_SCHEME_NAME"

mkdir_p RESULTS_PATH
desc "Run a test given a UI Automation script"
task :test, :file do |t, args|
        file = args[:file]

    sh %{ xcodebuild \\
            -project "#{PROJECT_DIR}" \\
            -scheme "#{BUILD_SCHEME}" \\
            -configuration Release \\
            -sdk iphonesimulator \\
            CONFIGURATION_BUILD_DIR="#{BUILD_DIR}" \\
            TARGETED_DEVICE_FAMILY=1 \\
            build }
    sh %{ instruments \\
            -t "#{AUTOMATION_TEMPLATE}" \\
            -D "#{OUTPUT_TRACE_DOCUMENT}" \\
            "#{APP_BUNDLE}" \\
            -e UIARESULTSPATH "#{RESULTS_PATH}" \\
            -e UIASCRIPT file }
    puts "\nTest Passed"
    sh %{killall "iPhone Simulator" || true}
end
```

Figure 2.7: Sample of a Rake task for executing an instruments test.

A Rake task defined as in the example above could also be invoked by executing the command shown in Figure 2.8.

```
rake test["automation/TEST_SUITE.js"]
```

Figure 2.8: Sample of a command to invoke a defined Rake task.

A specific Rake file with a name different from the default one could also be specified by using --rakefile option as seen in Figure 2.9.

```
rake --rakefile RakefileSequential test["automation/TEST_SUITE.js"]
```

Figure 2.9: Sample of a command to invoke a Rake task with --rakefile.

Rake is used throughout this work for many testing scenarios. For instance, to sequentially execute a set of test cases on a single machine or to test a particular test case given a path to a UI Automation script in a form of a Rake task.

2.2.4 VIRTUALIZATION

Virtualization is the process of emulating real IT resources with virtual IT resources. Virtual machines created from the process act exactly like actual IT resources with their own guest operating systems, which are independent of the host operating system that created them. Virtual machines run through the virtualization (hypervisor) software, which in turn run on a physical host, virtual machines can be easily copies and moved to another virtualization host as needed [20].

There are some limitations on using the normal execution environment to execute UI Automation test cases. First of all, only one test script can be run at a time. Therefore, all test cases in a test suite have to be run sequentially in this environment. It is also not possible to have multiple instances of the simulators running on a single machine, or to have multiple instances of the debuggers to run the tests on multiple devices attached to the machine. This means that this execution process could not be sped up simply by having multiple separated processes to execute test cases in parallel on a single machine.

Many practitioners suggested the use of multiple virtual machines running simultaneously on a single machine to solve the problem that only one instance of UI Automation debugger and simulator can be run at a time. Having multiple virtual machines run on a single machine means that there are multiple debuggers to run the tests. With limited resources, virtualization technologies can increase the efficiency and better utilization of the available resources in a cost effective manner. However, due to a strict licensing policies stipulated in Apple's software license agreement under the Section 2B(iii) [3], only two additional virtual machines of Mac OS X could be run on a single Mac machine. This restrictive license limits how far hardware resources could be utilized under this environment.

Although using virtualization technologies increase the number of debuggers and simulators to run the test, this solution still requires manual test distribution, meaning test cases needs to be manually selected and distributed to those virtual machines and manually executed. A good system should not require its users to be responsible for test case distribution. This also could slow down the process if the size of the cluster becomes large enough [36]. Nevertheless, this work-around suggests a distributed execution environment where there can be multiple instances of debuggers running in a cluster of machines, while test cases also get automated distribution. Virtualization technologies were used to set up such a cluster in various configurations for the experiments described in Chapter 4.

2.3   HADOOP AND HadoopUnit

This section provides a brief overview of Hadoop, including the Hadoop Distributed File System (HDFS) and the Map Reduce (MR) programming model, and the HadoopUnit distributed test execution environment that is built upon the Hadoop platform.

2.3.1   HADOOP

Hadoop is a collection of open source tools, libraries, and methodologies designed to run on commodity hardware or in the cloud that collectively serve as a scalable platform for big data analysis [56]. Hadoop consists of two core components: the Hadoop Distributed File System and MapReduce. A set of machines running these two components is known as a Hadoop cluster. An individual machine in a Hadoop cluster is known as a node and a cluster can consist of just one node or as many as a thousand of these nodes.

HDFS

Data in a Hadoop cluster is stored in the Hadoop Distributed File System (HDFS). In HDFS, data files are split into blocks, usually of 64 MB or 128 MB each, which are significantly larger than a conventional file system that is usually has a block size of 64 KB. These blocks are distributed across many different machines in the Hadoop cluster. Hadoop was designed with an assumption that hardware will fail, so some of the blocks are replicated on other machines, meaning that there are more than one machine storing the same copy of these blocks. In case of hardware failure, the same block of data could still be extracted from the other nodes in the cluster.

The node in the cluster that stores these blocks of data files is called DataNode. There is a single master node in the cluster called the NameNode that controls and keeps track of the locations of all these data blocks stored in different DataNode in the cluster as well as which blocks compose a file stored in the HDFS. There is also one or more Secondary NameNodes that act as checkpoints for the NameNode. In the case where the NameNode fails, the NameNode can be restarted using a backup snapshot stored in the Secondary NameNode [65] (Figure 2.10).

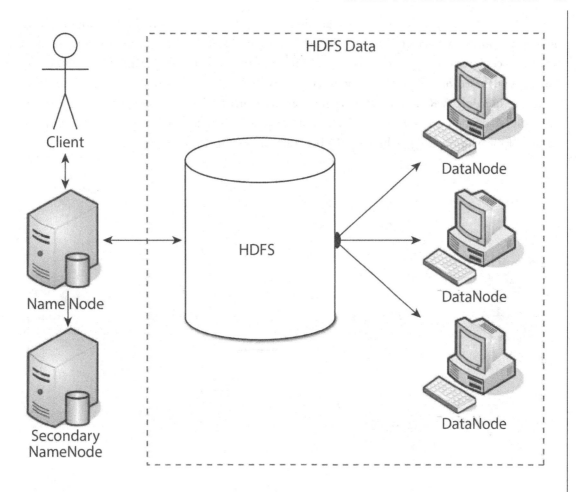

Figure 2.10: HDFS architecture.

MapReduce

Hadoop provides big data processing capability through the MapReduce (MR) programming model. The MapReduce programming model was built upon the concept that a computation is applied ("mapped") over a large number of records distributed all over the cluster to generate partial results that, in turn, are aggregated ("reduced") to produce the final solution [14]. The MapReduce programming model hides underlying execution details from users and provides automatic parallelization and distribution where developers can concentrate on writing data processing functions in a form of MapReduce jobs.

A MapReduce job consists of two functions: a Map function and a Reduce function. A Map function processes a given split of data derived from a block of data in a form of key/value pairs and produces intermediary output, also as a set of key/value pairs. The shuffle and sort works in the background, based on the key from the intermediary output, and feeds the output as an input to a Reduce function. A Reduce function aggregates the values of the processed intermediary output from the Map functions based on the key part of the given key/value pairs and provides the ultimate results of the MapReduce job. Figure 2.11 illustrates how MapReduce works.

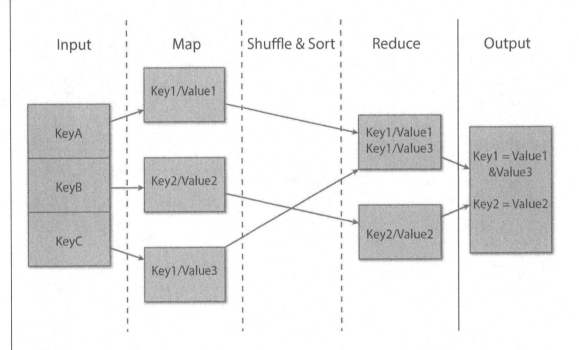

Figure 2.11: Overview of how MapReduce works.

MapReduce jobs are controlled by a software process running on the master NameNode called the JobTracker. Clients submit a MapReduce job through the JobTracker, which the Job-Tracker then assigns Map and Reduce tasks to, using the other DataNodes that stored a block of data to be processed in the cluster. These DataNodes each run a software process called the Task-Tracker. The TaskTracker receives Map or Reduce tasks from the JobTracker and instantiates the given task on the node and report progress back to the JobTracker. In case of task failures or in case when a particular task take unusually long, the tasks are restarted on other available nodes in the cluster automatically in the background by the JobTracker. Under speculative execution, if there is a free node available in the cluster, a redundant task will be assigned to it and the result will be collected from the node that finish executing the task first.

2.3.2   HadoopUnit

HadoopUnit is a distributed execution environment proposed by Parveen and Tilley [46] that is tailored specifically for executing JUnit test cases concurrently in the cloud. HadoopUnit was originally developed to address the need of mitigating the lengthy test execution of the Hadoop production code by using the Hadoop platform itself, because the method of test execution at that time took very long times to run and could not provide timely feedbacks to the developers. Hadoop was created using Java and its test cases are in the form of JUnit, hence HadoopUnit was originally created specifically for executing JUnit test cases [60].

HadoopUnit Architecture

HadoopUnit was built upon Hadoop with an adapted concept that a test suite is composed of a large set of test cases. Test cases are distributed all over machines in a cluster and the test execution is considered computing or processing upon them. The execution of test cases generates partial test results, which are part of the test suite. The final result is aggregated from those partial test results from the execution of test cases that are parts of a complete test suite.

HadoopUnit provides the infrastructure for JUnit test cases distribution and execution over a cluster of machines. HadoopUnit hides all the complexity of the cluster management and maintenance, job scheduling, resource allocation, and fault tolerance from the testers, thus letting them focus on what is important to them: testing.

As shown in Figure 2.12, there are three core functional components in HadoopUnit.

1. Test Case Extraction: This component of HadoopUnit is responsible for gathering all the test cases to be executed from the test suite and generating a test case list in the form of line-delimited string with each line composed of test case name and test execution command pair, a key/value pair format suitable to be processed by Hadoop. The original version of HadoopUnit used Ant, a Java build tool, as a component to provide this functionality.

2. The Map Function: This component of HadoopUnit is a Hadoop Map function responsible for receiving and processing given test case name/test execution command pairs by using the test case name as a key and executing the corresponding command as a separate process. The Map function produces intermediary results in a form of a test case name/test result pair, a key/value pair format suitable to be processed by the next component, the Reduce function.

3. The Reduce Function: This component of HadoopUnit is a Hadoop Reduce function responsible for collecting the intermediary test results from the Map function, aggregating into one report, and placing it in the HDFS where testers can extract the report to their machine.

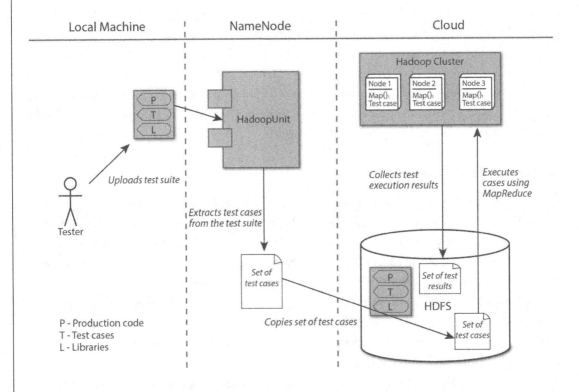

Figure 2.12: Overall architecture of HadoopUnit.

Testing with HadoopUnit

Testing with HadoopUnit begins with testers creating a HadoopUnit project and uploading the test suite to the file system where all the production code (a.k.a. the system under test), test cases for testing the production code, and dependent libraries were already in place in the HadoopUnit cluster. The test suite is then extracted and split into multiple sets of test cases that are distributed to machines in the Hadoop cluster. Each node executes the given set of test cases and returns back the test result. All the test results of all the execution of test cases are then being merged and reported back to the tester.

In keeping with HadoopUnit's goal of reducing the cost of running large regression test suites, it adopted the retest-all approach to software testing. While selective re-test techniques such as test case selection (test only what have been changed), test case prioritization (test the one with higher priority earlier), and test suite reduction (eliminate test cases from the test suite) can reduce the cost of regression testing, and indeed they can also be used in conjunction with HadoopUnit, they also reduce the chances that regression bugs can be revealed. Some set of test cases that could potentially expose those bugs might be eliminated in the reduction process [23, 50]. The retest-all approach is the simplest approach that avoids such errors and was selected by HadoopUnit; HadoopUnit shifts the focus to making the testing faster.

To be able to be executed concurrently under a distributed environment such as HadoopUnit, a set of test cases must be independent from one another. They cannot be executed at the same time without violating temporal dependencies if these test cases are co-dependent. The tests should not affect each other and the order of execution should also not affect the final test result. This means that the test should be self-contained, isolated, and fully functional. Although having this requirement in a set of test cases means that it can take longer to execute each test case, such isolated tests are valuable and provide higher quality feedback to the testers. This is because the failed test cases are not caused by any other dependent test cases in the test suite, which would require testers to further interpret the meaning of each failed test case [60, 7].

Requiring manual test case distribution not only slows down the whole testing process but also demands a substantial number of tasks be done by the testers themselves. After finishing the execution, testers also have to gather the results from the different machines to which they manually distributed the tests. This can be overwhelming, especially if the cluster size becomes large. HadoopUnit provides automated and transparent test case distribution, which is an ability that HadoopUnit inherited from Hadoop: it can push input data into nodes in the cluster automatically as well as to collect results and report back when finished. Test load distribution also is handled automatically by the underlying Hadoop platform for load balancing of the cluster. This is because if some of test cases take longer than another, they are distributed to other available nodes in the cluster automatically [9]. This approach of HadoopUnit better utilizes the available resources and prevents bottlenecks for rapid test execution.

CHAPTER 3

Using UI Automation with HadoopUnit

In migrating from a traditional execution environment to a distributed execution environment like HadoopUnit, careful, up-front planning needs to be taken, especially when planning to automate the testing process. Simply using a good set of tools does not guarantee the success of the project. This chapter provides information on UI Automation test suites, HadoopUnit customization, and using HadoopUnit as a platform for performing GUI testing of iOS applications.

3.1   UI AUTOMATION TEST SUITES

In order to migrate from a traditional GUI testing environment to the HadoopUnit distributed execution environment, the System Under Test (SUT) and its test suite must be prepared so that an execution of different test cases can be performed independently in parallel in the new environment. It is recommended practice that a test suite be developed in conjunction with the application so that the test could be run frequently as the development processes [48]. It is worth noting that an automated GUI testing project should be treated as a genuine programming project [33].

Once an application is released, it may be used or serviced for years. During the course of time, an application may inevitably need to be fixed (patches or bug fixes), changed (improvement or enhancements, or environmental changes such as operating system upgrade), or extended (adding new functionality to the application) [8]. The tests, once written, must also be maintained along with the application. Using capture/replay tools helps testers getting up and run to build a test suite quickly. However, for the test suite to be maintainable, some considerations on the test case design are needed so that the test cases will have such qualities. The test suite needs to evolve along with the software project [12].

The UI of an application will change throughout the life cycle of software development, especially in agile methodologies where changes in requirements are welcome, as long as the changes satisfy the users. Test cases need to be designed to be able to cope with this flexibility so the costs of such modification to the test case are kept to the minimum. The goal is to reduce the amount of work required to update the test scripts [39, 12].

3.1.1   TEST CASE DESIGN

Every feature of an application should be defined and written as a function within a test suite where test cases that need to test these features will need to access these functions. In other words, instead of having test cases to contain scripts for tapping a particular button to achieve some particular tasks, a function for such tasks should be created to wrap the underlying interactions with the UI around inside the function. This practice hides the low-level implementation of how the feature could be accessed. When the UI of the feature changes, the function that accesses that feature is the only place that needs to be updated, instead of every test case that tests the feature. It's also a good idea to pull out common scripts and create utility functions that other test cases can use, so when more functionality is added to the function, such as additional logging capabilities, then all test cases that use this function gain the additional functionality automatically [31, 35].

Jonathan Penn suggested a novel technique for designing UI Automation scripts for better maintenance by representing each screen of an application as a screen object with necessary methods to access GUI elements displayed on the screen [47]. The test cases that need to have access to the screens can import the script containing the screen object and perform testing on GUI elements as necessary with the provided methods. This way, if the user interface of the screen was changed, only the script representing the screen needs to be updated, instead of every place that needs to access that changed user interface, thus reducing the ripple effect from UI changes to test scripts.

For HadoopUnit, while the dependency within the same test case is acceptable, different test cases in a test suite needs to be designed in such a way that they can be executed independently—meaning the order of execution of test cases should not affect the test result. Because the test case execution will be executed on different nodes on a cluster, such dependency can cause problems in test execution.

This characteristic is also called test case contamination: the test case t_A should not contaminate the test case t_B no matter which one got executed first [60, 36]. A test case should set up the necessary data and other information required for testing without requiring other test cases (self-contained).

There is also another benefit of keeping test cases independent: a test case could be updated, improved, or refactored without affecting the other test cases [27]. Since each test case is independent of each other, when a test case fails, testers know exactly which test case to look for rather than requiring the tester to analyze which other test cases also depend on the failed test case.

3.1.2   TEST CASE ANALYSIS

Test case analysis is necessary for an application with existing UI Automation test suites in order to migrate to HadoopUnit environment. For example, the existing test suites might contain all test cases in one UI Automation script file. Distributing test cases in such a test suite having only one master UI Automation script file is not possible because it requires a mechanism to selectively execute test cases inside the test suite. This is due to the limitation of UI Automation that a parameter cannot be sent to the UI Automation engine through the command line terminal to selectively execute a specific test case within a test suite. Such types of test suites require test case analysis to break the test suite down into several files of test cases in order to prepare the test suites to run on HadoopUnit distributed environment.

As mentioned in the previous section, the UI Automation test script file representing each test case should be self-contained and must be independent of one another. It can contain many tests but it should not require the other test file to work. If there are such dependencies between them, the tester should consider grouping them together into one script file.

The result of test case analysis is a set of UI Automation test script files, each representing a single test case. Each file should be executable with Instruments.

3.2   HadoopUnit CUSTOMIZATION

The original version of HadoopUnit was designed specifically to execute JUnit test cases for Java applications, hence, it operates under a different environment from that of iOS applications. Some customization had to be made to set up HadoopUnit to prepare it for GUI testing of iOS applications. This section describes changes to the operational environment, gathering test results, and the resultant HadoopUnit architecture.

3.2.1   OPERATIONAL ENVIRONMENT

In order to migrate GUI testing of iOS application from a traditional environment towards a distributed execution environment like HadoopUnit, a Hadoop cluster needs to be set up in a way that each node is be able to compile, build, and run iOS applications. This requires each node in the cluster to have the Xcode, Instruments, iOS simulator, iOS SDK, as well as the Command line tool for OS X installed on the node. These tools operate under the Mac OS X operating system—Apple's proprietary operating system based on UNIX. This environmental requirement leads to several limitations and considerations that HadoopUnit has to accommodate.

Only one test case can be run at a time during test execution. This limitation means that regardless of the number of test cases in a test suite, they have to be executed sequentially [4]. This is not a problem for a distributed execution environment like HadoopUnit since there are many different nodes of machines available in the cluster to execute the test case concurrently. However, HadoopUnit needs to be configured to execute only one Map function at a time to accommodate this limitation, rather than the default value of two.

Only one instance of the iOS simulator can be run at a time. This is a limitation of the iOS simulator, a tool for running the application under test itself. This limitation is similar to the test case execution limitation, which means that another instance of iOS simulator cannot be started to run another set of tests on a single machine. In this case, the number of nodes available in the cluster directly correlates to the number of simulators available for test case execution.

As described in Section 2.2.4, due to licensing restrictions, only two additional instances of an OS X virtual machines can be run and they have to run on Apple's hardware only [3]. This requirement limits the size of the Hadoop cluster that could be constructed given available hardware resources. This restrictive requirement also makes it difficult to move this test execution environment to the cloud. Mac OS X isn't available as an instance type on major cloud providers such as Amazon's EC2 or Microsoft's Azure. This means that a Hadoop cluster of Mac OS X machines for testing iOS applications currently needs to be constructed and used internally only.

HadoopUnit as a platform for distributed test execution needs to account for these limitations of the operational environment so that it can support GUI testing of iOS applications. More detailed information on customizing a Hadoop cluster can be found in Appendix A.

3.2.2   TEST RESULTS

The Instruments tool, as mentioned in Chapter 2, generates test results not only as stream of log messages displayed on command-line terminal, but also records everything in a trace document. This trace document can be later opened with the Instruments GUI tool for identifying where and when the application under test went wrong. Hence, this document is very important and valuable for follow-up testing and inspections. The original version of Hadoop records test results only from the command-line output stream. Generating test reports from only this source make it difficult for testers to trace the behaviors of the application under test. HadoopUnit retrieves this document after finish executing each test.

The original version of HadoopUnit, after the test case extraction process, receives a test case list in a form of line-delimited string of test case name and test execution command key/value pairs. The location of the trace document is specified within the executing instruments command. However, to analyze the given execution command to extract the location of the trace document would make HadoopUnit less generic. Therefore, an additional parameter is used to direct HadoopUnit to retrieve additional documents on each execution node that perform Map functions and put them into the HDFS, so that testers can conveniently download the documents to their machine for further investigation.

The cost of moving files in and out of the HDFS is costly and time-consuming because it requires many I/O and network operations between many different nodes in the cluster [38]. Consequently, this version of HadoopUnit has been designed in such a way that only when a test case fails will HadoopUnit be triggered to retrieve trace documents from the node. In other words, if the whole test suite passes, no document is needed to be retrieved. This decision is reasonable because the purpose of regression testing is to find tests that fail; test cases that pass are not as interesting during regression testing. Given the above considerations, the test case list for HadoopUnit is updated in the form of a line-delimited string of test case name and test execution command pairs with the location of the trace document to be retrieved—an extended key/value triplet.

3.2.3   REVISED ARCHITECTURE

After undergoing the necessary customizations to run iOS GUI tests, HadoopUnit still has the same three core components as the original version of HadoopUnit, with slightly additional functionality and some customized behaviors according to its limitations and consideration: test case extraction, the Map function, and the Reduce function.

Test case extraction is an important component of HadoopUnit: it generates a test cases list in the form of line-delimited strings of test case name and execution commands. In this updated version, this list also includes the location of the output trace document to be retrieved by HadoopUnit after execution terminates. For the sake of simplicity, however, Ant was not used during the course of the experiments in this work. Instead, the test case extraction process was performed manually using a simple script to construct the list beforehand. The execution command is in a form of a simple Rake task that wraps around all the important classpaths, the location of the application bundle, the location to store the result trace document, as well as other scripts to take care of underlying work, such as building the application under test before testing or closing the simulator after finish executing each test. The code for the Rake task used, as well as the format of the list, is found in Appendix B.

The Map component of HadoopUnit is still responsible for actually executing each test case, given line-delimited strings of test case names, test execution commands, and the location of trace documents to be retrieved by HadoopUnit. The given test case name is used as a unique key for each test execution. Test results are merged by the Reduce function based on this key, so if there are lines with the same test case name, their test results will be merged together into one line on the report. For this experiment, the test case name is merely the name of the UI Automation script file used in the execution. In the case when a test execution failed, meaning that the application did not behave in an expected way as provided in the test script, the location of the trace document for that line will be used by HadoopUnit to retrieve and store the trace document in the HDFS, where testers can easily download the document for further investigation.

The Reduce component of HadoopUnit simply merges all the test results into one report using keys, which are the test case names of the test execution, and placed into the HDFS.

Figure 3.1 illustrates the overall architecture of HadoopUnit after being customized for iOS app GUI testing. The process begins with clients uploading a test case list to be used by HadoopUnit to the HDFS. The test case list is in a form of line-delimited string of test case names and test execution commands, with the location of the trace document to be gathered by HadoopUnit. Each node in the cluster is equipped with all necessary tools required for GUI testing of iOS applications, such as Xcode, Instruments, and iOS simulator, as well as an Xcode project of an application under test. The test case list is split and forms a set of map tasks to be distributed. After finishing executing the test, the output is stored in the HDFS as well as any additional documents specified with the test execution command. Clients then download the test report from the HDFS to their local machine. If some nodes failed, the map task given to the node will be automatically restarted on the other node in the cluster.

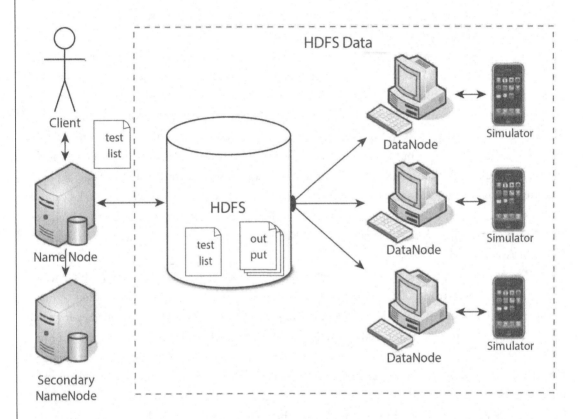

Figure 3.1: Architecture of HadoopUnit for GUI testing of iOS applications.

3.3   USING HadoopUnit

The following explains how the customized HadoopUnit is used as a platform for distributed GUI testing of iOS applications. A discussion of the test case list, Rake, and the test execution process is provided. This includes all steps necessary, from uploading the iOS test cases to interpreting the test results.

3.3.1   TEST CASE LIST

The list of test cases to be executed is the most important piece of information to be provided to HadoopUnit. It is a text file consisting of line-delimited text of test case names and their corresponding test execution command. It is provided to HadoopUnit in this form because it is the simplest that could be easily handled by Hadoop.

A sample of a test case list is shown in Figure 3.2. Each line of text in a test case list consists of three components separated by the symbol "@": test case name, test execution command, and the location of the trace document.

```
testcase_01.js @ rake test["automation/testcase_01.js"] @ test_results/
testcase_02.js @ rake test["automation/testcase_02.js"] @ test_results/
testcase_03.js @ rake test["automation/testcase_03.js"] @ test_results/
testcase_04.js @ rake test["automation/testcase_04.js"] @ test_results/
testcase_05.js @ rake test["automation/testcase_05.js"] @ test_results/
```

Figure 3.2: Sample of a test case list.

A test case name is used as a key to uniquely identify each test execution by the Map function. It is used as a key to merge together test results into one report by the Reduce function. There can be multiple executions of different commands for a single test case name, meaning that there can be multiple lines in the list with the same test case name. In this case, after finish executing all the test cases, each test execution will be merged into one line by the Reduce function. When a test case fails, this test case name is used as a name of directory to store the trace document in the HDFS.

A test execution command is the directive to execute a test case. The command will be executed as an external process by the Map function of HadoopUnit. This component can be in a form of a simple Instruments command or any type of executable command. The execution commands demonstrated in Figure 3.2 are in a form of executable Rake tasks receiving a path to a UI Automation test script. More information about Rake is explained in the next section.

The location of the trace document is an extra component that is added to each line of text in this version of HadoopUnit to accommodate the way Instruments treats the test results of each test case execution. This is the directory that Instruments uses to store a trace document for the test execution. When a test case fails, HadoopUnit uses this information to retrieve the trace document of the test execution and copy it into the HDFS.

The original version of HadoopUnit utilizes Ant, a build tool for Java projects, to extract test cases to be executed from a test suite, as well as other information needed to run the test cases such as necessary classpaths or libraries required to build and run a particular project. This test case list, however, can be constructed manually using a simple text editor or using a simple script to generate the list programmatically; no special tool is necessary.

Once the test case list has been created, it needs to be uploaded to the HDFS. Its path in the HDFS will be used as one of a parameter in the command to initiate a Hadoop job. The command to initiate a Hadoop job will be explained in detail in the following sections.

3.3.2   RAKE

As explained in Section 2.2.3, Rake is a Ruby build tool similar to Make. It makes the commands to execute each test case much simpler. Rake was chosen because it is already available in Mac OS X without any further installation required by the users. In this book, a Rake task is created and used to wrap around long execution of Instruments commands, and to provide additional functionality before and after executing each test case. An executable Rake task is used as a command in the test case list instead of a regular Instruments command to run the test.

In addition to executing shell scripts, the following functionality is utilized in HadoopUnit.

- Define common paths in variables: The important paths necessary to execute UI Automation scripts are, for example, the location of the application bundle, automation template, location to store test results, as well as location of Xcode project file to be able for a Rake task to compile and build the project. These paths can be stored in variables and used by commands within Rake tasks.

- Perform operations on directories: Some operations on directories are necessary in order to ensure a success in executing of a UI Automation test. For instance, the directory to store the test result needs to be created before executing each test case, a command to create a directory can be executed before executing an instruments command. Another directory operation before executing each test case is used to clean up the test results of the previous test before each execution. This ensures that even in the case of corrupted trace documents from the previous test will not affect the current test execution.

- Receive additional parameters from the command-line terminal: Additional parameters can be sent to Rake to customize the behavior of its execution. For example, to execute a test given a path to a UI Automation script, to execute the test on a different device family (iPhone or iPad), or to execute the test using a different scheme.

- Kill process: After each test case has finished execution, the simulator process is killed to reset all settings and configuration data that might have left from the test so that it is ready for the next test case to be executed. This task is to ensure that different test cases execute independently.

There are two main Rake tasks created for two different scenarios: a Rake task to sequentially execute test cases on a single machine (which provides a baseline for comparing results using HadoopUnit) and a Rake task that receives a path to a UI Automation script as a parameter to execute different test case in a test suite. The Rake file is part of the Xcode project, so that a Rake task can access the actual test script on each node in the cluster. The full source code (with comments) of the Rake task for executing UI Automation test can be found in Appendix B.

3.3.3   TEST EXECUTION

This section outlines the steps necessary to execute GUI test cases for iOS apps using the customized version of HadoopUnit.

Uploading an iOS Project to Each Node in the Cluster

Each node in the cluster is required to have a working iOS project in the node. This can be done using any kind of network file transfer mechanism, such as File Transfer Protocol (FTP) or Server Message Block (SMB). If each node is configured to enable File Sharing with Apple Filing Protocol (AFP), file sharing in a Mac OS X network with the Finder is made easier. This enables the project files to be easily distributed to each node over the network.

Uploading a Test Case List to the HDFS

The test case list previously created needs to be made available in the HDFS. It must be uploaded only through the Hadoop file system command on the master NameNode. Otherwise, the test case list won't be accessible by Hadoop to perform its operations. An example of Hadoop file system command to transfer a file from a local machine to the HDFS is shown in Figure 3.3.

```
        hadoop fs -copyFromLocal <LOCAL_DIRECTORY>/testlist.txt
<HDFS_DIRECTORY>/testlist.txt
```

Figure 3.3: Sample of command to transfer a file to the HDFS.

Starting Test Executions with HadoopUnit

To initiate a Hadoop job to start a test execution process with HadoopUnit, after the iOS project of an application under test and the test case list have been properly uploaded, the Hadoop jar command as seen in Figure 3.4 is executed through the command-line terminal. The HadoopUnit. jar used by the command should be located on local machine used to execute the command.

```
hadoop jar <LOCAL_DIRECTORY>/HadoopUnit.jar TestDriver
<HDFS_DIRECTORY>/testlist.txt <HDFS_DIRECTORY>/output
```

Figure 3.4: The Hadoop command to initiate test execution with HadoopUnit.

There are four important arguments that need to be specified with this command.

1. A JAR file that contain a MapReduce code. The HadoopUnit.jar is actually a MapReduce program packaged in a form of JAR file.

2. The class name in the JAR file that contains a method to drive the MapReduce code, which in this case is TestDriver.

3. The input file, which is the test case lists previously created and uploaded to the HDFS. The path of the input file is the path of the file located in the HDFS. This file will be accessed by Hadoop to perform the tasks necessary to split and distribute test cases to different nodes in the cluster.

4. The output directory. This is a directory where Hadoop stores the results of execution. This directory is also where HadoopUnit stores the trace documents of failed test cases. The directory, however, needs to be empty in the HDFS or the Hadoop job can't be started.

After the command has been initiated, a Hadoop job is created. The input file will be accessed by Hadoop to perform tasks necessary to split the test cases list into what Hadoop calls input splits. Those input splits are then distributed to several nodes in the cluster automatically. The number and size of each split are determined by the value of **mapred.max.split.size**, which is pre-configured in the Hadoop configuration file. Detailed information on configuring HadoopUnit are found in Appendix A. After executing every test case, the test results are merged by the Reduce function and placed in the output directory specified in the command.

Gathering Test Results

After all the test cases have finished executing, the test results are gathered from the HDFS using simple a Hadoop file system command, as shown in Figure 3.5. The test results from Hadoop as well as a trace document of failure test cases are downloaded from the directory in the HDFS to the specified directory on a local machine.

```
hadoop fs -copyToLocal <HDFS_DIRECTORY>/output <LOCAL_DIRECTOTY>/output
```

Figure 3.5: The Hadoop command to download files from the HDFS.

Test Results Analysis

Test results from test executions are stored in a text file with a name given by Hadoop similar to "part-r-00000." The content of the file is a list of test case names and the result of each test—whether its execution passed or failed. If a test case was labeled as failed, its corresponding trace document also be downloaded and stored in a directory named by its test case name.

The results from these trace documents need to be investigated and follow-up testing performed on the failed test cases as necessary, perhaps by executing individual test cases separately on a local machine. There is also a chance of a false alarm where there was an error from the test execution resulting from the testing tool instead of from the misbehavior of an application under test.

At the time of writing (with Xcode 5.1), it has currently been marked as known issues in its release note. After the failure has been identified and fixed accordingly, an updated source code for the system under test will need to be re-uploaded as well an updated sets of test cases (if they exist) to each node in the Hadoop cluster so that another test run can be done.

CHAPTER 4

Rapid GUI Testing of iOS Apps

The traditional method of GUI testing with UI Automation can only execute each test sequentially on a single machine. This requires a considerable amount of time for the test execution process to complete if there are a lot of test cases in the test suite. This chapter describes how rapid GUI testing of iOS apps is done using HadoopUnit. A series of experiments was carried out with various testing configurations (e.g., cluster size). The performance of HadoopUnit in each of these contexts is analyzed. Threats to validity of the experiments are discussed.

4.1 EXPERIMENTS

Several experiments were set up to perform GUI testing of iOS apps with different scenarios using HadoopUnit. These experiments included creating a baseline for comparison by running sets of test cases with a different number of tests (100, 250, 500, and 1,000) sequentially on a single machine. The number of test cases was chosen to simulate different sizes and complexity of a representative system under test. A large and complex application would contain many test cases, whereas a smaller one would contain a fairly smaller set of test cases. The same sets of test cases were then run concurrently on a small 2-node and 4-node cluster with HadoopUnit and their execution times compared with the baseline.

A Hadoop cluster of two nodes and four nodes were set up for the experiments. Each node was running as a virtual machine of Mac OS X 10.8.5 Mountain Lion allocated with two processor cores and 3 GB of RAM on VMware Fusion 6.0.2. The system under test, which was developed specifically for these experiments, was a simple money management app where users could add, edit, or remove their daily money related activities. The application is shown in Figure 4.1.

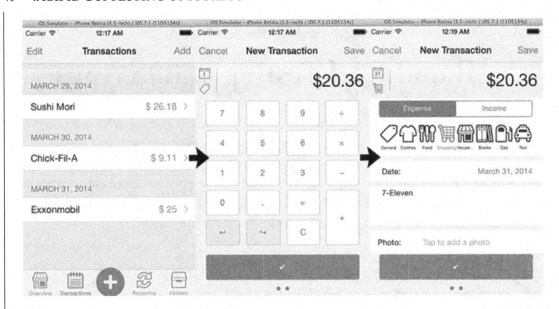

Figure 4.1: A sample screenshot of the system under test.

As mentioned above, the experiments were conducted using four different test suites consisting of 100, 250, 500, and 1,000 test cases running at separate times. Test cases in each test suite were the result of duplicating the same test cases performing the same set of tasks. Each test case consists of three tests shown in Figure 4.2.

```
test("Creating a 'new' transaction", function() {
        var main = MainScreen;
        var newtrans = NewTransactionScreen;

        main.gotoTransactionsScreen();
        main.addNewTransaction();

        newtrans.addAmount(123);
        newtrans.addNote('Test Case 1');

        newtrans.submitTransaction();
        main.gotoOverviewScreen();
});

test("Seeing a transaction's details", function() {
        var main = MainScreen;
        var trans = TransactionsScreen;
        var detail = TransactionDetailScreen;

        main.gotoTransactionsScreen();

        trans.tapTransaction("Test Case 1, $ 123");
        trans.target().captureScreenWithName("Test Case 1");
        detail.goBack();
});

test("Removing a transaction", function() {
        var main = MainScreen;
        var trans = TransactionsScreen;

        main.gotoTransactionsScreen();

        trans.removeTransaction("Test Case 1, $ 123");
        trans.assertNoTransaction("Test Case 1, $ 123");
});
```

Figure 4.2: Sample of a test case.

The first test was to add a new expense transaction with a test case name as its description and a sample amount of expense and then return to the main list after adding a new transaction. The second test was to assert that a new transaction is actually being added to the list by tapping the added transaction in the list and checking the detailed information of the transaction before returning to the main list. The last test was to remove the added transaction from the list and assert that the transaction was successfully removed from the list. These three tasks compose a single test case in this experiment and are duplicated to create a test suite. Each test case create a transaction with a different note, for example, a test case number 2 would add a note as "Test Case 2" while a test case number 100 would add a note as "Test Case 100" and so on, for monitoring purposes.

Each test case used in this experiment takes, on average, 32 s to complete its execution of the three tasks as mentioned earlier. The execution includes the time it takes for the simulator to launch the application, perform transitions between different views, enter text using a virtual keyboard, and perform animations during transaction removal, and the times it takes for UI Automation itself to check assertions.

The following sections provide detailed information on the experiments performed to determine the effectiveness of HadoopUnit as a platform for GUI testing of iOS applications, as well as the performance gain from the migration.

4.1.1   EXPERIMENT I

The first experiment was conducted to provide a baseline for the rest of the experiments by executing each test suite sequentially on a single machine without using HadoopUnit. A Rake task was created to run each test case, one after another. The tasks being performed before and after each test execution, as mentioned in the Chapter 3, are the same as the tasks used in another Rake task for running with HadoopUnit. The execution time was recorded and reported after executing every test case in the test suite. The code of this Rake task is shown in Figure 4.3.

```
decs "Run a set of test cases sequentially"
task "test" do
  start_time = Time.now

  automate "automation/test_case_1.js"
automate "automation/test_case_2.js"
automate "automation/test_case_3.js"
automate "automation/test_case_4.js"
automate "automation/test_case_5.js"

total = Time.now - start_time
put "Finish executing 5 tests, took #{total} seconds"
  end
```

Figure 4.3: Code for executing test cases sequentially with a Rake task.

The experiments were run using 100, 250, 500, and 1,000 test cases to simulate the different size and complexity of the application under test. The execution time was then analyzed to determine whether it scaled linearly as the number of test cases increases.

Results

Each set of experiments was conducted 10 times and the average execution time from the experiments was computed and recorded. The execution times for this experiment are shown in Table 4.1 and Figure 4.4.

Table 4.1: Sequential execution time on a single machine (in seconds)				
	100 Test Cases	**250 Test Cases**	**500 Test Cases**	**1,000 Test Cases**
Sequential Runtime (s)	3,018.81	7,788.68	16,505.96	35,352.39

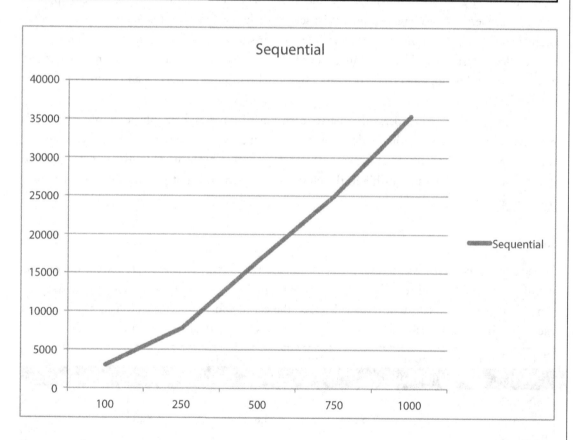

Figure 4.4: Sequential execution time on a single machine.

The 100 test cases took 3,018.81 s or 50 min and 18 s. The 250 test cases took 7,788.68 s or 2 hr, 9 min, and 48 s. The 500 test cases took 16,505.96 s or 4 hr, 35 min, and 5 s. The 1,000 test cases took 35,352.39 s or 9 hr, 49 min, and 12 s.

Analysis

The results for this baseline experiment indicate that the execution time increased approximately linearly as the number of test cases in a test suite increased. It took excessively long time (almost 10 hr) to run the set of 1,000 test cases sequentially on a single machine. This clearly indicates the needs for a better execution environment for this kind of testing, which is the main purpose for this book in the first place.

4.1.2   EXPERIMENT II

The second experiment was conducted with the same set of test suites but with the addition of HadoopUnit. A 2-node cluster running HadoopUnit was set up for GUI testing of iOS applications. This experiment was created to reflect a scenario where a group of developers added a modest single machine capable of running two virtual nodes or simply added two physical machines to set up a Hadoop cluster with HadoopUnit. A set of test case lists was created with the number of test cases corresponding for each test, 100, 250, 500, and 1,000 lines in the list and supplied as input to HadoopUnit.

Each Map task is a process that parses each line of input and executes each test case. Each map task runs on a node in the Hadoop cluster. Since there can be only one test case execute at a time in one node, the number of concurrent test execution is determined by the number of node capable of execute the test available in the cluster. For which node got which split of the test case list is determined solely by Hadoop.

Results

Each set of experiments was conducted 10 times and the average execution time from the experiments was computed and recorded. Execution times of the experiments were reported by the job's progress that Hadoop constantly sends back to the command-line terminal during the course of execution. The execution times of the experiments are shown in Table 4.2 and Figure 4.5.

Table 4.2: Concurrent execution time on a 2-node cluster (in seconds)			
	100 Test Cases	**250 Test Cases**	**500 Test Cases**	**1,000 Test Cases**
2-Node Concurrent Runtime (s)	1,523	3,876	7,820	16,023

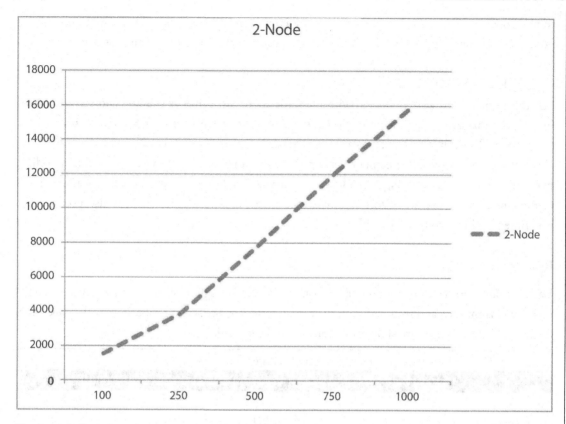

Figure 4.5: Concurrent execution time on a 2-nodes cluster.

The 100 test cases took 1,523 s or 25 min and 23 s. The 250 test cases took 3,876 s or 1 hr, 4 min and 36 s. The 500 test cases took 7,820 s or 2 hr, 10 min, and 20 s. The 1,000 test cases took 16,023 s or 4 hr, 27 min, and 3 s. There was, however, an observable delay of about 5–15 s at the beginning of each execution before a new Map process be spawned on each node in the cluster. This delay is further discussed in the next section.

Analysis

The result from the third experiment shows an approximate 2x reduction in test execution time when compared with the baseline case of a sequential test case execution on a single machine. The execution time of a test suite with 100 test cases was reduced from 50 min and 18 s down to 25 min and 23 s. The execution time of a test suite with 250 test cases was reduced from 2 hr, 9 min, and 48 s down to 1 hr, 4 min, and 36 s. The execution time of a test suite with 500 test cases was reduced from 4 hr, 35 min, and 5 s down to 2 hr, 10 min, and 20 s. Finally, the execution time of a test suite with 1,000 test cases was reduced from 9 hr, 49 min, and 12 s down to 4 hr, 27 min, and 3 s. The result shows that a performance improvement of approximately 2x can be achieved by adding a Hadoop cluster running with HadoopUnit, even if the size of the cluster is as small as just two nodes.

4.1.3   EXPERIMENT III

The third experiment was conducted with the same set of test suites as the first and second experiments, but this time with a 4-node cluster running HadoopUnit set up for GUI testing of iOS applications. This experiment reflects a scenario where a group of developers set up a larger Hadoop cluster, keeping the license agreement that Apple has for their operating system in mind, by adding two more virtual nodes running on another machine or simply adding two more physical machines to the cluster. The same set of test case lists used in the second experiment was also used in this experiment as input to the HadoopUnit.

Results

Each set of experiments was conducted 10 times and the average execution time from the experiments was computed and recorded with the same method as the second experiment. The execution times of the experiments are shown in Table 4.3 and Figure 4.6.

Table 4.3: Concurrent execution time on a 4-node cluster (in seconds)				
	100 Test Cases	**250 Test Cases**	**500 Test Cases**	**1,000 Test Cases**
4-Node Concurrent Runtime (s)	866	2,185	4,332	8,452

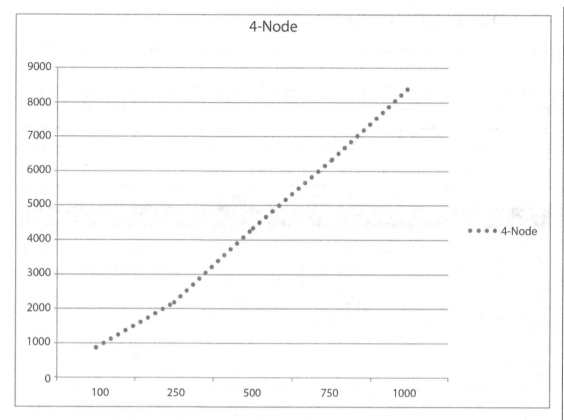

Figure 4.6: Concurrent execution time on a 4-node cluster.

The 100 test cases took 866 s or 14 min and 26 s. The 250 test cases took 2,185 s or 36 min and 25 s. The 500 test cases took 4,332 s or 1 hr, 12 min, and 12 s. The 1,000 test cases took 8,452 s or 2 hr, 20 min, and 52 s. There were also observable delays at the beginning of each test execution as found in the second experiment; this is discussed in the next section.

Analysis

The result from the third experiment shows an approximate 3.5x to 4x reduction in test execution time when compared with the baseline case of a sequential test case execution on a single machine. The execution time of a test suite with 100 test cases was reduced from 50 min and 18 s down to 14 min and 26 s. The 250 test cases was reduced from 2 hr, 9 min, and 48 s down to 36 min and 25 s. The 500 test cases was reduced from 4 hr, 35 min, and 5 s down to 1 hr, 12 min, and 12 s. The 1,000 test cases was reduced from 9 hr, 49 min, and 12 s down to 2 hr, 20 min, and 52 s. The results from this experiment are promising and suggest that a performance gain proportional to the number of nodes in the cluster can be gained even from a small-scale implementation of HadoopUnit.

4.2   DISCUSSION OF RESULTS

The results from the second and the third experiment show that given a small scale Hadoop cluster of four nodes or even as small as two nodes, there can be a performance gain closely related in a linear fashion to the number of nodes in the cluster. The amount of time it took to execute the test suite increased linearly as the number of test cases in a test suite grew—the same behavior for all three experiments. Table 4.4 and Figure 4.7 provide a comparison of execution times for the three experiments. Additionally, Table 4.5 highlights the improvement factor of HadoopUnit when compared with the traditional sequential execution on a single machine.

Table 4.4: Performance comparisons of the three experiments (in seconds)

	100 Test Cases	250 Test Cases	500 Test Cases	1,000 Test Cases
Sequential Runtime (s)	3,018.81	7,788.68	16,505.96	35,352.39
2-Node Concurrent Runtime (s)	1,523	3,876	7,820	16,023
4-Node Concurrent Runtime (s)	866	2,185	4,332	8,452

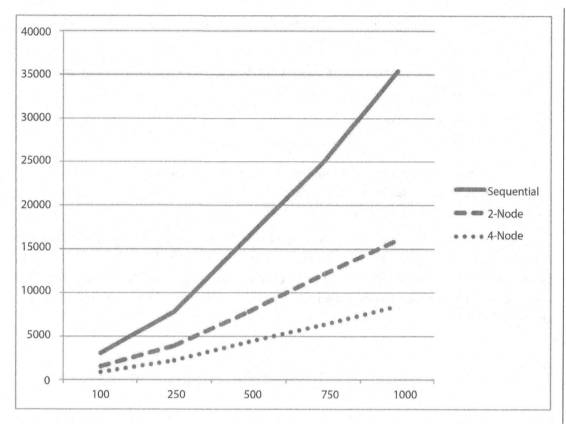

Figure 4.7: Execution time comparison of the three experiments.

Table 4.5: Performance factors over sequential execution				
	100 Test Cases	**250 Test Cases**	**500 Test Cases**	**1,000 Test Cases**
2-Node Concurrent Runtime (s)	1.98	2.00	2.11	2.20
4-Node Concurrent Runtime (s)	3.48	3.56	3.81	4.18

When compared to the baseline sequential test execution on a single machine, distributed GUI testing with HadoopUnit showed an approximate 2x performance gained when executed with a 2-node cluster and an approximate 3.5x to 4x performance gained when executed with a 4-node cluster. The following summarizes the execution time reduction when executing GUI testing of iOS apps with HadoopUnit when comparing with traditional test execution on a single machine:

- 50 min of execution time could be reduced to about 25 min on a 2-node cluster and about 14 min on a 4-node cluster;

- 2 hr of execution time could be reduced to about 1 hr and 4 min on a 2-Node cluster and about 36 min on a 4-node cluster;

- 4 hr of execution time could be reduced to about 2 hr and 9 min on a 2-node cluster and about 1 hr and 12 min on a 4-node cluster; and

- an excessively long 9 hr and 49 min of execution time could be reduced to 4 hr and 27 min on a 2-node cluster and to just 2 hr and 20 min when comparing with such a baseline on a single machine on a 4-node cluster.

As seen from the experimental results, HadoopUnit outperforms the traditional test execution that can only execute each test case sequentially on a single machine in every case, because more than one test case is executed in parallel, thus accelerating the process. Based on these experiments, the total test execution time can be roughly approximated with the equation shown in Figure 4.8.

$$T = \left(\frac{\Sigma test}{\Sigma node} \times \bar{t} \right)$$

Figure 4.8: Total test execution time approximation equation.

Where:

- T is the total execution time

- $\Sigma test$ is the total number of test cases in a test suite

- $\Sigma node$ is the total number of nodes in the cluster

- $\bar{t}$ is an average execution time for each test case

Because each test case can only be executed sequentially on a single machine, in a normal case, the total amount of time it takes to execute the whole test suite can be computed from the total number of test divided by the total number of nodes available in the cluster for executing the test and multiply by the average execution time of a test case.

$$T = t_{longest}$$

Figure 4.9: Ideal case for test execution time approximation equation.

The ideal case for a distributed execution environment, as shown in Figure 4.9, is the situation where the total number of test is equal to the total number of nodes. Then the time it takes for a whole test suite to finish is the execution time of the longest test case in the test suite (excluding overhead). By this reasoning, it is not recommended to have a single test case that is excessively longer than the other test cases in the test suite. Ideally, they should be split evenly in terms of time needed to complete: the speed of executing a test suite is determined by the time it takes for the longest test case to finish executing.

As evidenced from the experimental results, there are cases where the performance improvement did not match the number of nodes in the cluster as expected. This is because of the overhead introduced by Hadoop. Hadoop sacrifices some runtime performance at the expense of providing desirable underlying features, such as automatic input splitting, task scheduling and distribution, and fault tolerance. This frees developers from worrying about these factors while developing MapReduce applications [15]. Similarly, it frees testers from the same concerns when using HadoopUnit.

The processes that contribute overhead to the execution performance with Hadoop are the transformation of the input file input key/value pairs, the statistical tasks throughout the execution for decisions on job scheduling, and the processing logic of the MapReduce programming model. This processing is broken down into several phases: map, shuffle and sort, and reduce. All of these mechanisms contribute additional overhead to the overall execution time with Hadoop.

There were also noticeable delays at the beginning of each execution when testing with HadoopUnit. It was observed during the course of the second experiment that it took 5–15 s at the beginning of each test execution before a new process could be spawned on each node in the cluster. This is an overhead time introduced by Hadoop to set up tasks. This is not the case for sequential test execution on a single machine, where the test execution begins right away when the Rake task is invoked from the command line.

It is stated in the Hadoop documentation that it can take some time for each Hadoop Map task set up to complete. The documentation further suggests that the maps should take at least a minute to execute for large jobs [67, 55]. This statement implies that the more Map tasks, the more the set up times there will be for a single Hadoop job. Configuring the right number of Map tasks for a Hadoop job can be non-trivial. It depends on many parameters and configuration of the Hadoop environment, such as the number of nodes available in the cluster and the size of the input split used by Hadoop. The number of Map tasks for each job is determined by the size of the input file, which in these experiments is the test case list provided to HadoopUnit. The size of the input test case list derives from the length of each line of text in the test case list, which depends on the length of the test case name, the test execution command, and the trace document directory name, which is composed of a single line-delimited string.

Hadoop splits the given input file by determining the size of the HDFS block, so the split size is simply the size of the HDFS block [65]. However, the split size can also be controlled by modifying the value of the **mapred.max.split.size** parameter. Normally, the split size for each execution is set as default to the maximum value that could be represented by a Java long type and will have an effects only when the value is less than the HDFS block size. In these experiments, the parameter **mapred.max.split.size** was set to 1,000 bytes, since the test case list is significantly smaller than the default block of 64 MB in HDFS. Setting this parameter to a value less than the block size forces the split size to be smaller than a block and increases the number of Map tasks.

The Rake task used in the test case list to execute the test is also significantly smaller than a typical Instruments command to execute UI Automation. This makes the size of the test case list file smaller than usual. Under a circumstance where the Rake task is not used, this parameter should be further adjusted by increasing the split size. Increasing the split size results in fewer Map tasks per job and more test execution per split. However, there are trade-offs to consider when setting this parameter. If the value is too small, the test execution wastes start-up overhead and results in an inefficient shuffle and sort phase. If the value is too big, Hadoop can't provide enough parallelism to efficiently utilize the cluster to execute the given Hadoop job [10]. There is also a chance that some nodes in the cluster might fail. If such a circumstance arises, a Map task given to the node will be restarted on another node in the cluster. If the split size is large, the more execution time is wasted with the node, which might impact the overall performance of the test execution.

The following suggestions provide guidance to customize Hadoop in different situations.

- If the amount of test cases in the test suite is small, resulting in a small test case list, it is suggested to consider reducing the split size to be smaller than the size of the test case list. This will force concurrency in the test execution. Otherwise, Hadoop won't even bother distributing the Map tasks to the other nodes in the cluster, making a test execution with HadoopUnit ineffective due to the introduced overheads.

- If the number of nodes in the cluster increases, it may be desirable to consider reducing the split size appropriately to add the degree of concurrency to the test execution to take advantage of the additional IT resources. This practice might, however, increase the amount of overhead to the overall test execution, as there are more Map tasks to set up per Hadoop job. On other hand, if the number of nodes in the cluster is small, it might be beneficial to increase the split size to reduce the number of Map tasks, so the overhead incurred by them will be reduced. The configurations used in these experiments can be used for such a scenario.

- If the average execution time of the test cases in the test suite is small, consider increasing the split size to give more Map tasks to be executed per node. This will reduce the latency affect of Hadoop's task set up and introduced overhead as mentioned earlier.

For these experiments, each test case took around 32 s to finish executing its set of tests. The results from the experiments have shown the delay and the introduced overhead to be insignificant compared to the performance gain realized by migrating the traditional test execution to HadoopUnit.

4.3   THREATS TO VALIDITY

Some decisions regarding the experiments were made as this work progressed and could threaten the validity of the test result data and conclusions drawn from the experiments. Three such concerns are discussed in this section: test suites, Hadoop optimization, and network issues.

4.3.1   TEST SUITES

For the experiments performed, the test suites were built by duplicating the same test case to create a test suite with 100, 250, 500, and 1,000 test cases. Each test case performs exactly the same set of tasks with slightly different data to indicate a test case name currently being performed on the node, for monitoring its progress during execution.

The goal was to distribute loads equally on every node in the cluster to prove the effectiveness of each node being added to the cluster. Each node was expected to get the same amount of work and to get fully utilized throughout the test execution. Additionally, the execution time could be directly compared after the test execution.

However, using this practice might not reflect real-life scenarios where each test case runs at a variable amount of time rather than the execution times that are similar to each other. For example, there might be some test cases that run excessively longer than the others. There is also a type of test case where networking or location services are involved. This could affect how Hadoop distributes the workloads while making it harder to conclude its effectiveness on test executions. Using a controlled set of test cases removes this factor that might influence the performance of the system.

4.3.2   HADOOP OPTIMIZATION

Tuning the performance of Hadoop for a particular type of execution is subtle and is not a simple task. It involves with many different combinations of parameters available for administrators to be configured [29]. A modification of one parameter could result in Hadoop behaving in a subtly different manner that is hard to monitor. Some parameters are used as just a hint and it's totally up to Hadoop to decide how it should behave given a particular situation. This is the trade-off of the simplification that Hadoop provides for developers to implement a distributed execution system [38, 15].

The focus of this work is to prove the usefulness of HadoopUnit in reducing the execution time of GUI testing of iOS applications using a small-scale cluster. By this reason, only a few critical parameters that could help accomplish this goal were selected, which are mapred.tasktracker. map.tasks.maximum and mapred.max.split.size. There are other Hadoop parameters that could have been considered to further optimize the performance of the test execution. This form of Hadoop optimization is left for future work.

4.3.3 NETWORK ISSUES

A private network was used in the experiments to run a Hadoop cluster; the network was dedicated for use in the experiment only. Each node in the cluster was used solely for executing the tests. However, under a real production network used by many other individuals, the network traffic of the shared network might become a parameter that users who would like to adopt this system need to consider, since Hadoop relies heavily on communications between nodes in the cluster.

CHAPTER 5

Summary

Mobile devices are fast becoming the dominant computing platform worldwide. The apps running on these devices do everything from games and entertainment to supporting online finance to storing personal health records. This increased reliance on smartphone apps has increased the need for quality assurance, and a primary way of achieving this is through software testing—particularly with the GUI that users interact with every day.

However, GUI testing of smartphone apps on popular platforms like iOS has become very challenging. The time needed to perform thorough testing of a complex user interface has become so long that it often precludes running large test suites on a regular basis. A better solution is needed to address this problem.

This book describes how a customized version of HadoopUnit can be used to improve the testing process by reducing the amount of time needed to execute system-level GUI test cases with UI Automation for iOS apps. The time reduction is realized by leveraging the underlying capabilities of the Hadoop platform to distribute test cases to computing nodes on a cluster of machines, which execute the tests concurrently.

With faster test execution, developers and testers no longer have to wait excessively long periods of time for the testing process to complete in order to receive feedback from large regression test suites. This means they can build and test their code more frequently, thus improving the overall quality of the apps they are developing.

5.1 SUMMARY OF RESULTS

Several experiments were conducted to simulate various scenarios where a small-scale implementation of HadoopUnit could be used in a software project. The set up included the following three configurations:

- a single machine, sequential execution;

- a 2-node cluster with HadoopUnit, concurrent execution; and

- a 4-node cluster with HadoopUnit, concurrent execution.

The first configuration provides a baseline execution time for HadoopUnit, while the other two configurations were chosen to evaluate the effectiveness of HadoopUnit with a small cluster of two and four nodes, respectively. Comparing the result of normal sequential execution on a single machine to using HadoopUnit for concurrent execution, a nearly linear speedup was recorded for both 2-node and 4-node cluster configurations.

The results indicate that with minimal investment the testing process can be significantly improved using HadoopUnit. The results also suggest that the use of HadoopUnit can be greatly effective for agile methodologies, since testing is performed frequently throughout the life cycle (e.g., at the end of each iteration).

The results of this work can be summarized according to two criteria: how well the original research objectives were satisfied, and what new contributions this work has made to the software testing community.

5.1.1 RESEARCH OBJECTIVES

The primary objective of this work was to reduce the time it takes to perform system-level GUI testing for apps on the iOS platform. The approach used to meet this objective was to migrate testing from a traditional sequential execution environment on a single machine to HadoopUnit, a distributed execution environment running on a cluster of machines, where test cases are executed concurrently. The results described above show that this primary research objective was satisfied.

The secondary objective was to prove the effectiveness of a test execution after migrating to a distributed execution environment on a small-scale cluster, as opposed to a large-scale infrastructure of hundreds or thousands of nodes in the cloud. The rationale behind this objective is that not every software group has access to large amounts of IT resources all of the time. The results described above for 2-node and 4-node clusters show that this secondary objective was also satisfied.

5.1.2 RESEARCH CONTRIBUTIONS

The primary research contribution of this work was demonstrating the viability and efficacy of using HadoopUnit for system-level GUI testing of iOS apps. HadoopUnit has already proven its worth in past work related to regression testing of JUnit test cases. This research extended and customized the HadoopUnit platform to work equally well for mobile application testing.

The secondary research contribution of this research was demonstrating that HadoopUnit was valuable even on small 2-node and 4-node clusters for GUI testing of iOS apps. This modest infrastructure is accessible to almost all developers and testers, which means they can adopt HadoopUnit to realize the quality benefits of rapidly testing their mobile apps. The experiments indicated that reducing test execution time in a nearly linear fashion was achievable, even with the hardware, software, and licensing restrictions imposed by Apple and/or the UI Automation framework. In addition to a faster test execution, HadoopUnit also provides testers with several other benefits, such as automated test case distribution, test result gathering, and test load distribution.

5.2   FUTURE WORK

In addition to the benefits that HadoopUnit provides to an iOS software development project, there are several areas where the work of this book with HadoopUnit could be improved. For example, the process and workflow presented is performed in a semi-automated fashion, involving several manual processes to drive the test execution. It could be better if HadoopUnit was combined with a continuous integration server, like Hudson or Jenkin, which would further streamline other aspects of the software development lifecycle.

For example, whenever a developer commits a code to the server, the server automatically checks for any syntactical errors and starts rebuilding the project. If the build was a success, scripts on the server then could be started to upload the Xcode project and the source code to every pre-configured node in the clusters. Any updated test scripts for automation would also be uploaded along with the project. After finishing the uploading, each node then automatically starts building the project within the node and installs the app in the simulator. When finished, each node reports back to the server. The server updates the test case list to be used by HadoopUnit and start the distributed test execution with HadoopUnit on every node in the cluster. After finishing execution, test results are reported to the server while any trace document of the failure test case is made accessible to the team. All of these processes can happen in the background without requiring any user-interaction.

It might also be useful to extend this work to support the other major mobile platforms, such as Android or Windows Phone, so that HadoopUnit could help improve the quality of mobile applications in these other markets. Certainly the Android platform, which enjoys the largest market share of mobile devices on the planet, would benefit tremendously.

5.3   CONCLUDING REMARKS

We now live in a "Post-PC" era with the proliferation of mobile devices [1]. The speed of mobile application development and the quality of the apps are now more important than ever. Software testing is one way to help improve the app quality. If slow test execution times impede properly performing software testing, HadoopUnit is a viable option that addresses this challenge in a low-cost and effective manner. In the future, a crashing app won't simply be a case of gaming annoyance—it could be a matter of life and death.

Appendix A
Setting up a HadoopUnit Cluster on Mac OS X

PREREQUISITES

OPERATING SYSTEM

Each node that constitutes a HadoopUnit Cluster was operated primarily under the Mac OS X 10.8.5 Mountain Lion run under VMware Fusion Professional 6.0.2.

XCODE

Xcode must be installed on each node in a cluster in order to compile and build iOS working projects and to run the test via Instruments, which is also bundled with Xcode. At the time of writing, Xcode version 5.0.2 (5A3005) was the latest version available, but due to the instability issue with the currently available versions of Xcode, Xcode version 5.1 Beta 4 (5B90f) was primarily used to perform the experiments described in this book.

The following command is used to switch between versions of Xcode installed on a single machine.

```
sudo xcode-select -switch /Applications/Xcode51-Beta4.app
```

To check the current version of Xcode that the machine is working on, use the following command:

```
xcode-select -print-path
```

or

```
xcodebuild -version
```

COMMAND LINE TOOLS FOR XCODE

Command Line Tools for Xcode is required to enable a command-line workflow with UI Automation, including compiling, building, and running the UI Automation testing through the command-line interface. Command Line Tools for Xcode can be obtained from the Apple Developer website or directly inside of Xcode under the Download Components Section. Command Line Tools (OS X Mountain Lion) for Xcode—October 2013 was used in this book.

unix_instruments

The instruments command, at the time of writing, returns the same status code no matter whether the test case passes or fails. Unix_instruments is a wrapper script that was used to detect an error in execution by eavesdropping on the output from the instruments commands and return status code appropriately. For this work, this command is executed instead of normal instruments command. (https://gist.github.com/jonathanpenn/1402258)

JAVA

Java must be installed to run Hadoop on each node in the cluster. A working installation of Java SE 6 version 1.60_65 for OS X Mountain Lion was used. (http://support.apple.com/kb/dl1572)

SSH

Hadoop relies on SSH to communicate between nodes in the cluster and to perform cluster-wide operations. In order to work seamlessly, SSH should be configured to allow keyless/passwordless login for users from machines in the Hadoop cluster.

On Mac OS X, we first need to enable Remote Logins by System Preferences/Sharing/Remote Logins. Also note the Computer Name that will be used as a HOST_NAME during set up.

Then set up a RSA public/private key pair to be able to ssh into the node. On each node (both master and slave), type the following commands to generate RSA key pair of the node:

```
ssh-keygen -t rsa -P "" -f $HOME/.ssh/id_rsa
```

The private key is stored in the file specified by the –f option, in this case $HOME/.ssh/id_rsa. The public key is stored in the file with the same name but with a .pub extension appended, in this case $HOME/.ssh/id_rsa.pub.

Make sure that the public key is authorized by copying the public key into $HOME/.ssh/authorized_keys using the following command:

```
cat $HOME/.ssh/id_rsa.pub >> $HOME/.ssh/authorized_keys
```

Make sure that host names in the cluster are configured correctly in the host file at /etc/host so that each node can communicate by its name rather than its IP-address. Then ssh to localhost machine and the actual host names to make sure that ssh is working correctly. Both the master node and the slave node must be able to ssh to each other. This step will also add the hosts' fingerprint into the known_hosts file. Note that the HOST_NAME of a Mac OS X machine can be found in System Preferences/Sharing.

```
ssh localhost

ssh USERNAME@HOST_NAME
```

Finally, distribute the public key of the master node to all slave nodes in the cluster. On the master node, execute the following command to append the public key to a remote host that act as a slave node:

```
cat $HOME/.ssh/id_rsa.pub | ssh USERNAME@HOST_NAME 'cat >> $HOME/.ssh/
authorized_keys'
```

HADOOP INSTALLATION

At the time of writing, Hadoop version 1.2.1 is the stable version and was used in this work. The current release of Hadoop can be downloaded from the Apache Hadoop Releases website (http://hadoop.apache.org/releases.html), then the downloaded package can be unpacked under the location of choice. A quick installation guide for Hadoop can be found in Hadoop Wiki (http://wiki.apache.org/hadoop/QuickStart).

However, the easy way for Mac OS X to install Hadoop is through the Homebrew tool, a package management manager for OS X (http://brew.sh). The tool can be obtained on the fly using the following command through the command-line terminal:

```
ruby -e "$(curl -fsSL https://raw.github.com/Homebrew/homebrew/go/in-
stall)"
```

After Homebrew has been installed on the machine, Hadoop can be installed with Homebrew simply by using the following command:

```
brew install hadoop
```

Or, use the following command to check the available versions first in order to install a specific version of Hadoop:

```
brew search hadoop

brew install homebrew/versions/hadoop121
```

Homebrew will install Hadoop in /usr/local/Cellar/hadoop/<version> and will also set $JAVA_HOME to /usr/libexec/java_home.

HADOOP CONFIGURATION

There are six configuration files that need to be customized. They are located in /usr/local/Cellar/hadoop/1.2.1/libexec/conf.

1. hadoop-env.sh

2. core-site.xml

3. hdfs-site.xml

4. mapred-site.xml

5. masters

6. slaves

hadoop-env.sh

This file sets environment variables that are used in the scripts to run Hadoop. Homebrew does all this work during installation, but there is an issue in Mac OS X Lion and Mountain Lion that requires some configuration in this file to resolve the issue (https://issues.apache.org/jira/browse/HADOOP-7489). Add the following line to the file:

```
export HADOOP_OPTS="-Djava.security.krb5.realm=-Djava.security.krb5.kdc="
```

core-site.xml

This file sets the configuration settings for Hadoop Core, such as the I/O settings of the nodes. This file must be configured on every node in the cluster:

```
<configuration>
        <property>
                <name>fs.default.name</name>
                <value>hdfs://[MASTER_HOST_NAME]:9000</value>
        </property>
        <property>
                <name>hadoop.tmp.dir</name>
                <value>/tmp/hadoop-${user.name}</value>
        </property>
</configuration>
```

The `fs.default.name` must point to the master node only with the correct port that the master node is listening to.

The `hadoop.tmp.dir` is the directory Hadoop uses for write working temporary files.

hdfs-site.xml

This file controls the configuration for Hadoop Distributed File System process, the name-node, the secondary name-node, and the data-nodes.

```
<configuration>
        <property>
                <name>dfs.replication</name>
                <value>4</value>
        </property>
        <property>
                <name>dfs.permissions</name>
                <value>false</value>
        </property>
</configuration>
```

The `dfs.replication` controls the number of replications when a file is created in HDFS. In this book, our data is the test case list and we want our data to be replicated on every node in the cluster, so this value should be set to the value equal to the number of nodes in the cluster.

The `dfs.permissions` is set to false to avoid permission issues during the execution in the experiments. This value means that any user can do anything to HDFS but since the users need to be able to login to the Mac OS X in the first place, turning this off seems to be reasonable to eliminate any possible problems.

mapred-site.xml

This file controls the configuration of MapReduce process, the job tracker, and the tasktrackers.

```xml
<configuration>
        <property>
                <name>mapred.job.tracker</name>
                <value>[MASTER_HOST_NAME]:9001</value>
        </property>
        <property>
                <name>mapred.tasktracker.map.tasks.maximum</name>
                <value>1</value>
        </property>
        <property>
                <name>mapred.tasktracker.reduce.tasks.maximum</name>
                <value>1</value>
        </property>
        <property>
                <name>mapred.max.split.size</name>
                <value>1000</value>
        </property>
</configuration>
```

Again, the `mapred.job.tracker` must point to the master node only with the correct port since only the master node runs the job tracker in Hadoop cluster.

The `mapred.tasktracker.map.tasks.maximum` controls the number of map tasks running per node. Since the limitation of iOS simulator that we can only have 1 simulator run at a time, we cannot have more than one map task that run the test on a single node. The default value of this property is set to 2; we have to explicitly set the `mapred.tasktracker.map.tasks.maximum` to the value of 1.

The `mapred.tasktracker.reduce.tasks.maximum` controls the number of reduce tasks running per node. Since the job of the Reduce function is trivial, the mapred.tasktracker.reduce.tasks.maximum is set to the value of 1 as well.

The `mapred.max.split.size` is directly correlated to how Hadoop splits and distributes the input file throughout the HDFS. Since the input file is just a list of test execution commands, it is not big (probably in a unit of a few MB rather than GB or TB). Each test execution is considered computing extensive and time-consuming. This value is set so small that it will force Hadoop to distribute the job to the other nodes as well, otherwise Hadoop won't even bother distributing the jobs and just run on a single node because Hadoop will consider given input test case list to be a small data. For the experiments, this value was set to 1000 or 1 KB, around 8–10 lines per split.

masters

The masters file is a list of machine's host names or IP-address that each run a secondary name-node (not the machine that runs as the master name-node but the secondary name-node—although it could be the master node as well). In the experiments, the master node not only acts as the name-node but also as a secondary name-node too. The masters file contains just the following line, which is set only on the master node:

```
MASTER_USERNAME@MASTER_HOST_NAME
```

slaves

The slaves file is a list of machine's host names or IP-address that each run a data-node and Task-Tracker in the cluster. The master node can also act as a data-node so the master node can appear in this list as well. This file needs only be set on the master node. The number of nodes in the cluster can be easily added or removed by modifying this file.

```
MASTER_USERNAME@MASTER_HOST_NAME
SLAVE_USERNAME_1@SLAVE_HOST_NAME_1
SLAVE_USERNAME_2@SLAVE_HOST_NAME_2
SLAVE_USERNAME_3@SLAVE_HOST_NAME_3
```

TEST RUNNING HADOOP

Before starting the Hadoop cluster, initialize the HDFS by using the following command:

```
hadoop namenode -format
```

Then to start the Hadoop cluster, execute the following command:

```
/usr/local/Cellar/hadoop/1.2.1/libexec/bin/start-all.sh
```

Determine whether the Hadoop cluster is working correctly by running some samples provided with Hadoop installation, for example, with the following command. The command should give us the result text "Estimated value of Pi is 3.14800000000000000000."

```
hadoop jar /usr/local/Cellar/hadoop/1.2.1/libexec/hadoop-examples-*.jar
pi 10 100
```

To stop the Hadoop cluster, use the following commands:

```
/usr/local/Cellar/hadoop/1.2.1/libexec/bin/stop-all.sh
```

TEST EXECUTION WITH HadoopUnit

The HadoopUnit source code is packaged in .jar format. There are three classes: TestDriver.java that contains the main method and drive this MapReduce application; TestMapper that executes the

Map function to perform actual test execution; and TestReducer that executes the Reduce function to gather those test results from the Map function and combines them into one report.

To execute testing with HadoopUnit, upload the test case list to the HDFS with the following command:

```
hadoop fs -copyFromLocal <LOCAL_DIRECTORY>/testlist.txt <HDFS_DIREC-
TORY>/testlist.txt
```

Then execute the following command to start the process:

```
hadoop jar <LOCAL_DIRECTORY>/HadoopUnit.jar TestDriver <HDFS_DIREC-
TORY>/testlist.txt <HDFS_DIRECTORY>/output
```

The HadoopUnit.jar should reside locally in a client machine that wants to run the tests. There is no need to upload this package into the HDFS because Hadoop will try to find the jar file on the local machine only. The TestDriver is the name of the class that contains the main method that drives the MapReduce application. The testlist.txt is the test case list that was previously uploaded to the HDFS. The output is the directory in the HDFS to which HadoopUnit will store the test results.

After finished executing the test, the test results can be gathered from the HDFS by using the following command:

```
hadoop fs -copyToLocal <HDFS_DIRECTORY>/output <LOCAL_DIRECTOTY>/output
```

The previous test results could be removed from the HDFS with the following command before the next test could be run (or other directory in the HDFS can be specified):

```
hadoop fs -rmr <HDFS_DIRECTORY>/output
```

There is an additional step necessary to automate iOS GUI testing on an individual node in the cluster. When the testing with Instruments gets started, there will be a dialog asking for a permission to control another process. This security feature gets in the way of automating the process, as we need to manually acknowledge this dialog every time the test is run. To by pass this process, the following command needs to be executed on every node in the cluster:

```
sudo /usr/libexec/PlistBuddy \
    -c "Set rights:system.privilege.taskport:allow-root true" \
    /etc/authorization
```

The PlistBuddy command will set a specific key necessary to grant permission for Instruments to control other processes in the /etc/authorization file. The permission dialog may appear at first after executing the command, but the process will be on autopilot after that.

Note that in order for a machine to be able to run a test, first agree to the Xcode iOS license, which requires admin privileges. This is done simply by opening Xcode the first time and Xcode will prompt us with a dialog asking for an admin password. Also note that Developer Mode must be enabled on the machine running the test. This is done by opening an Xcode project and starting to build it. Xcode will prompt with another dialog asking to enable this mode. This is a necessary trivial step needed in order to get every ready for testing.

Appendix B
HadoopUnit Source Code for
iOS GUI Testing

TestDriver.java

```java
import org.apache.hadoop.fs.Path;
import org.apache.hadoop.io.Text;
import org.apache.hadoop.conf.Configuration;
import org.apache.hadoop.conf.Configured;
import org.apache.hadoop.util.Tool;
import org.apache.hadoop.util.ToolRunner;
import org.apache.hadoop.mapreduce.Job;
import org.apache.hadoop.mapreduce.lib.input.FileInputFormat;
import org.apache.hadoop.mapreduce.lib.output.FileOutputFormat;

public class TestDriver extends Configured implements Tool {

  // The Driver Class
  public int run(String[] args) throws Exception {
        String input, output;
        if (args.length == 2) {
                input = args[0];
                output = args[1];
        } else {
                System.err.println("Incorrect number of arguments.  Ex-
pected: input output");
                return -1;
        }

        Configuration conf = getConf();
        Job job = new Job(conf);
        job.setJarByClass(TestDriver.class);
        job.setJobName(this.getClass().getName());

        // Specify the input directory from which data will be read,
        // and the output directory to which output will be written.
        FileInputFormat.setInputPaths(job, new Path(input));
        FileOutputFormat.setOutputPath(job, new Path(output));
```

```java
// Specify which classes are to be the Mapper and Reducer
        job.setMapperClass(TestMapper.class);
        job.setReducerClass(TestReducer.class);

        // Specify the types of the intermediate
        // output key and value produced by the Mapper
        job.setMapOutputKeyClass(Text.class);
        job.setMapOutputValueClass(Text.class);


// Specify the types of the Reducer's output key and value.
        job.setOutputKeyClass(Text.class);
        job.setOutputValueClass(Text.class);

// Finally, run the job
        boolean success = job.waitForCompletion(true);
        return success ? 0 : 1;
    }


// The main method simply calls ToolRunner.run(),
// passing in the Driver class and the command-line arguments.
    public static void main(String[] args) throws Exception {
        TestDriver driver = new TestDriver();
        int exitCode = ToolRunner.run(driver, args);
        System.exit(exitCode);
    }
}
```

TestMapper.java

```java
import java.io.BufferedReader;
import java.io.BufferedWriter;
import java.io.IOException;
import java.io.InputStream;
import java.io.InputStreamReader;
import java.io.OutputStream;
import java.io.OutputStreamWriter;
import java.util.StringTokenizer;
import java.util.regex.Matcher;
import java.util.regex.Pattern;

import org.apache.hadoop.fs.FileSystem;
import org.apache.hadoop.fs.Path;
import org.apache.hadoop.io.LongWritable;
import org.apache.hadoop.io.Text;
import org.apache.hadoop.mapreduce.Mapper;
import org.apache.hadoop.mapreduce.lib.output.FileOutputFormat;
```

```java
public class TestMapper extends Mapper<LongWritable, Text, Text, Text>
{

  public void map(LongWritable key, Text value, Context context) throws
IOException, InterruptedException {

// Get the line
        String inputLine = value.toString();
        StringTokenizer stok = new StringTokenizer(inputLine, "@");

        String testName = stok.nextToken().trim();
        String command = stok.nextToken().trim();
        String resultPath = stok.nextToken().trim();

        // Regular Expression for separating string with white space
// but not that between single quotes
// (Process does not play well with a space)
// String with white space should be put inside single quote
        ProcessBuilder builder = null;
        Pattern p = Pattern.compile("(?<=\\s|^)(\'.*?\'|\\S*)
(?=$|\\s)");
        Matcher m = p.matcher(command);
        int index = 0;
        while (m.find()) {
                if (index == 0) {
                        builder = new ProcessBuilder(m.group(1));
                }
                else {
                        builder.command().add(m.group(1));
                }
                index++;
                }

        StringBuilder testResult = new StringBuilder();
        builder.redirectErrorStream(true);
        Process process = builder.start();

        OutputStream stdin = process.getOutputStream();
        InputStream stderr = process.getErrorStream();
        InputStream stdout = process.getInputStream();

        BufferedReader reader = new BufferedReader(new InputStream-
Reader(stdout));
        BufferedWriter writer = new BufferedWriter(new OutputStream-
Writer(stdin));

        String line = reader.readLine();
        while (line != null) {
                testResult.append(line);
                line = reader.readLine();
```

```
            }

        process.waitFor();
        int exitValue = process.exitValue();
        if (exitValue == 1) {
                testResult.insert(0, "Failed ");

                // If the test failed, copy the trace document to HDFS
                try {
// Put some wait time until instruments finish writing to the trace
file.
                        Thread.sleep(5000);

                        Path phdfs_input = new Path(FileOutputFormat.get-
OutputPath(context).toString().trim()+Path.SEPARATOR+testName.trim());
                        Path plocal_input = new Path(resultPath);
                        FileSystem fs = FileSystem.get(context.getConfigu-
ration());
                        fs.copyFromLocalFile(false, false, plocal_input,
phdfs_input);
                } catch (Exception e) {
                        e.printStackTrace();
                }
        }
        else {
                testResult.insert(0, "Passed ");
        }

        String errString = null;
        BufferedReader stdError = new BufferedReader(new InputStream-
Reader(stderr));
        while ((errString = stdError.readLine()) != null) {
                System.err.println(errString);
        }

        // Send test name and test results to reducer
        context.write(new Text(testName), new Text(testResult.to-
String()));
    }
}
```

TestReducer.java

```
import java.io.IOException;
import org.apache.hadoop.io.Text;
import org.apache.hadoop.mapreduce.Reducer;

public class TestReducer extends Reducer<Text, Text, Text, Text> {
```

```java
@Override
  public void reduce(Text key, Iterable<Text> values, Context context)
throws IOException, InterruptedException {

        StringBuilder testResult = new StringBuilder();
        for (Text value : values) {
                testResult.append(value.toString());
        }

        context.write(key, new Text(testResult.toString()));
  }
}
```

Rakefile

```ruby
BUILD_DIR               = "/tmp/ExpenseKit"
APP_BUNDLE              = "#{BUILD_DIR}/ExpenseKit.app"
AUTOMATION_TEMPLATE     = "ExpenseKit/automation/MyTemplate.tracetem-
plate"
RESULTS_PATH            = "ExpenseKit/automation_results"
OUTPUT_TRACE_DOCUMENT = "#{RESULTS_PATH}/Trace"
PROJECT_DIR             = "/ExpenseKit/ExpenseKit.xcodeproj"
BUILD_SCHEME            = "ExpenseKit"

# If the automation_results directory isn't there, Instruments com-
plains.
mkdir_p RESULTS_PATH

desc "Run appropriate tests for iPhone and iPad Simulators"
task :test, :file do |t, args|

  #receive an argument named file, access it using args[:argument_name]
  file = args[:file]

  sleep 3
  clean_results
  build "iphone"
  automate file
  reset_sim

  puts "\nWin condition acquired!"
end


#
# Composable steps
#

# Remove the automation_results directory and start fresh
```

```ruby
def clean_results
  rm_rf RESULTS_PATH
  mkdir_p RESULTS_PATH
end

def clean
  run_xcodebuild "clean"
end

def build type
  case type
  when "iphone"
    sdk = "iphonesimulator"
    fam = "1"
  when "ipad"
    sdk = "iphonesimulator"
    fam = "2"
  when "device"
    sdk = "iphoneos"
    fam = "1,2"
  else
    raise "Unknown build type: #{type}"
  end

  run_xcodebuild "build -sdk #{sdk} TARGETED_DEVICE_FAMILY=#{fam}"
end

def automate script
  #reset_sim

  if $is_testing_on_device
    device_arg = "-w #{connected_device_id}"
  end

  #env_vars = extract_environment_variables(script)

  sh %{
    ExpenseKit/bin/unix_instruments \\
      #{device_arg} \\
      -t "#{AUTOMATION_TEMPLATE}" \\
      -D "#{OUTPUT_TRACE_DOCUMENT}" \\
      "#{APP_BUNDLE}" \\
      -e UIARESULTSPATH "#{RESULTS_PATH}" \\
      -e UI_TESTS 1 \\
      -e UIASCRIPT "#{script}"
  }
end

def automate_normal script
```

```ruby
    #reset_sim

    if $is_testing_on_device
      device_arg = "-w #{connected_device_id}"
    end

    #env_vars = extract_environment_variables(script)

    sh %{
      instruments \\
        #{device_arg} \\
        -t "#{AUTOMATION_TEMPLATE}" \\
        -D "#{OUTPUT_TRACE_DOCUMENT}" \\
        "#{APP_BUNDLE}" \\
        -e UIARESULTSPATH "#{RESULTS_PATH}" \\
        -e UI_TESTS 1 \\
        -e UIASCRIPT "#{script}"
    }
end

def close_sim
  sh %{killall "iPhone Simulator" || true}
end

def reset_sim
  close_sim
  sim_root = "~/Library/Application Support/iPhone Simulator"
  rm_rf File.expand_path(sim_root)
end

#
# Utility Methods
#

def run_xcodebuild extra_args
  sh %{
    xcodebuild \\
      -project "#{PROJECT_DIR}" \\
      -scheme "#{BUILD_SCHEME}" \\
      -configuration Release \\
      CONFIGURATION_BUILD_DIR="#{BUILD_DIR}" \\
      #{extra_args}
  }
end

def extract_environment_variables script
  lines = File.readlines script
  arguments = []
```

```ruby
    lines.each do |line|
      line.match(%r{^// (.+)=(.+)$})
      if $1
        arguments << "-e " + $1 + " " + $2
      end
    end

    arguments.join(" ")
end

def ioreg_output
  `ioreg -w 0 -rc IOUSBDevice -k SupportsIPhoneOS`
end

def connected_device_is_ipad?
  !ioreg_output.match(/"USB Product Name" = "iPad"/).nil?
end

def connected_device_id
  ioreg_output.match(/"USB Serial Number" = "([A-z\d]+)"/) && $1
end

def install_on_device
  # I got fruitstrap originally from here:
  # https://github.com/ghughes/fruitstrap
  #
  # It's no longer supported and you might need to use a fork on
Github
  raise "No device connected" if !connected_device_id
  sh %{bin/fruitstrap -b #{APP_BUNDLE} -i #{connected_device_id}}
end
```

References

1. Appcelerator. (2013, November). Enterprise Mobile Application Development Platform | Appcelerator Inc. [Online]. http://www.appcelerator.com/enterprise/resource-center/research/appceleratoridc-q4-2013-mobile-developer-report/. 5, 57

2. Apple Inc. (2011, June) Apple Developer. "Automating User Interface Testing with Instruments." [Online]. http://developer.apple.com/devcenter/download.action?path=/wwdc_2010/wwdc_2010_video_assets__pdfs/306__automating_user_interface_testing_with_instruments.pdf. 2, 5

3. Apple Inc. (2012, August) Apple - Legal - Software. "Software License Agreement for OS X Mountain Lion." [Online]. http://images.apple.com/legal/sla/docs/OSX1082.pdf. 5, 19, 30

4. Apple Inc. (2013, November) iOS Developer Library. "Instruments User Guide: Automating UI Testing." [Online]. https://developer.apple.com/library/ios/documentation/DeveloperTools/Conceptual/InstrumentsUserGuide/UsingtheAutomationInstrument/UsingtheAutomationInstrument.html. 2, 3, 5, 12, 14, 29

5. Apple Inc. (2013, September) iOS Developer Library. "Cocoa Core Competencies: Objective-C." [Online]. https://developer.apple.com/library/ios/documentation/general/conceptual/DevPedia-CocoaCore/ObjectiveC.html. 5

6. Ashkenas, J. (2009, December) "CoffeeScript." [Online]. http://coffeescript.org. 5

7. Beck, K. *JUnit Pocket Guide*. O'Reilly Media, 2004. 25

8. Black, R.; van Veenendaal, E.; and Graham, D. *Foundations of Software Testing: ISTQB Certification* (3rd ed.) Cengage Learning, 2012. 2, 9, 11, 27

9. Bower, E. *Performance Analysis of a Distributed Execution Environment for JUnit Test Cases on a Small Cluster*. MSE Thesis, Florida Institute of Technology, Melbourne, FL, 2010. 4, 25

10. Cloudera. (2012, June). Cloudera. "Optimizing MapReduce Job Performance." [Online]. http://www.cloudera.com/content/cloudera/en/resources/library/presentation/hadoop-summit-2012-optimizing-mapreduce-job-performance-presentation-slides.html. 52

11. CloudMonkey. Mobile App Testing Tool. [Online]. http://www.cloudmonkeymobile.com/monkeytalk. 5

12. Collins, E.; Neto, A.; and de Lucena, V. "Strategies for Agile Software Testing Automation: An Industrial Experience." *Proceedings of the 36th IEEE International Conference on Computer Software and Applications Workshops* (COMPSACW 2012) , Izmir, Turkey, 2012, pp. 440–445. 27

13. Cunha, M.; Paiva, A.; Ferreira, H.; and Abreu, R. "PETTool: A Pattern-Based GUI Testing Tool." *Proceedings of the 2nd International Conference on Software Technology and Engineering* (ICSTE 2010), vol. 1, San Juan, PR, 2010, pp. V1-202–V1-206. 2

14. Dean, J. and Ghe*mawat, S. "MapReduce: A Flexible Data Processing Tool."* Communications of the ACM, vol. 53, 2010, pp. 72–77. DOI: 10.1145/1629175.1629198. 21

15. Ding, M. et al., "More Convenient More Overhead: The Performance Evaluation of Hadoop Streaming." *Proceedings of the 2011 ACM Symposium on Research in Applied Computation* (RACS '11). New York, NY, USA, 2011, pp. 307–313. DOI: 10.1145/2103380.2103444. 51, 54

16. Duarte, A. and Sávio de Oliveira, G. "A Framework for Automated Software Testing on the Cloud." *Proceedings of the 4th International Conference on Parallel and Distributed Computing, Applications and Technologies* (PDCAT'13), 2013. 6

17. Duarte, A.; Cirne, W.; Brasileiro, F.; Duarte, P.; and Machado, P. "Using the Computational Grid to Speed up Software Testing." *Proceedings of 19th Brazilian Symposium on Software Engineering*, 2005. 7

18. Duarte, A.; Wagner, G.; Brasileiro, F.; and Cirne, W. "Multi-Environment Software Testing on the Grid." *Proceedings of the 2006 Workshop on Parallel and Distributed Systems: Testing and Debugging*, New York, NY, USA, pp. 61–68. 7

19. Dutil A. et al. (2013, June) SourceForge.net. "iPhone Guitar." [Online]. http://sourceforge.net/apps/mediawiki/guitar/index.php?title=IPhone_Guitar. 5

20. Erl, T.; Mahmood, Z.; and Puttini, R. *Cloud Computing: Concepts, Technology & Architecture.* Prentice Hall, 2013. 19

21. Fowler, M. Xunit. [Online]. http://www.martinfowler.com/bliki/Xunit.html. 10

22. Gartner. (2013, September) "Gartner Says Mobile App Stores Will See Annual Downloads Reach 102 Billion in 2013." [Online]. http://www.gartner.com/newsroom/id/2592315. 1

23. Graves, T.; Harrold, M.-J.; Kim, J.; Porter, A.; and Rothermel, G. "An empirical study of regression test selection techniques." *ACM Transactions on Software Engineering and Methodology* (TOSEM), vol. 10, New York, NY, USA, 2001, pp. 184–208. DOI: 10.1145/367008.367020. 3, 25

24. Greiner, L. (2014, October) Financial Post. "IBM and SAP teaming up in the cloud." [Online]. http://business.financialpost.com/2014/10/27/ibm-and-sap-teaming-up-in-the-cloud/?__lsa=829c-dcc2. 2

25. Harrold, M.-J.; Gupta, R.; and Soffa, M.-L. "A Methodology for Controlling the Size of A Test Suite." *Proceedings of the Conference on Software Maintenance*, San Diego, CA, 1990, pp. 302–310. DOI: 10.1145/152388.152391. 3

26. Hellesøy, A. "Cucumber - Making BDD Fun". [Online]. http://cukes.info. 5

27. Holmes, A. and Kellogg, M. "Automating Functional Tests Using Selenium." *Agile Conference*, 2006, Minneapolis, MN, 2006, pp. 6–275. DOI: 10.1109/AGILE.2006.19. 28

28. IEEE. "IEEE Standard Glossary of Software Engineering Terminology." IEEE Std 610.12-1990, pp. 1–84, December 1990. DOI: 10.1109/IEEESTD.1990.101064. 11

29. Impetus Technologies. (2010, November) Hadoop toolkit. "Hadoop Performance Tuning." [Online]. https://code.google.com/p/hadoop-toolkit/downloads/detail?name=White%20paper-HadoopPerformanceTuning.pdf&can=2&q=. 53

30. Issa, A.; Sillito, J.; and Garousi, V. "Visual Testing of Graphical User Interface: an Exploratory Study Towards Systematic Definitions and Approaches." *Proceedings of the 14th IEEE International Symposium on Web Systems Evolution* (WSE 2012), Trento, Italy, 2012, pp. 11–15. DOI: 10.1109/WSE.2012.6320526. 12

31. Kaner, C. (1997, May). "Improving the Maintainability of Automated Test Suites." [Online]. http://kaner.com/pdfs/autosqa.pdf. 28

32. Kaner, C. (2002, September). "Paradigms of Black Box Software Testing." [Online]. http://kaner.com/pdfs/ParadigmsTutorial.pdf. 11

33. Kaner, C. (2002). "Avoiding Shelfware: A Managers' View of Automated GUI Testing." [Online]. http://www.kaner.com/pdfs/shelfwar.pdf. 27

34. Kaner, C. and Fiedler, L. "Foundations of Software Testing." [Online]. http://www.testingeducation.org/BBST/foundations/. 9

35. Kaner, C.; Bach, J.; and Pettichord, B. *Lesson Learned in Software Testing: A Context Driven Approach*. John Wiley & Sons, Inc., 2002. 28

36. Kapfhammer, G. "Automatically and Transparently Distributing the Execution of Regression Test Suites." *Proceedings of the 18th International Conference on Testing Computer Software*, Washington, DC, 2001. 3, 19, 28

37. Li, Y. and Wahl, N. "An Overview of Regression Testing." *ACM SIGSOFT Software Engineering Notes*, vol. 24(1), NY, USA, 1999, pp. 69–73. 11

38. Lin, X.; Meng, Z.; Xu, C.; and Wang, M. "A Practical Performance Model for Hadoop MapReduce." *Proceedings of the International Conference on Cluster Computing Workshops*, Beijing, China, 2012, pp. 231–239. 30, 54

39. Lowell, C. and Smith, J. "Successful Automation of GUI Driven Acceptance Testing." *Proceedings of the 4th International Conference on Extreme Programming and Agile Processes in Software Engineering*, vol. 2675, Genoa, Italy, 2003, pp. 331–333. DOI: 10.1007/3-540-44870-5_43. 3, 12, 27

40. Marchetto, A. and Ricca, F. "Transforming a Java Application in an Equivalent Web-Services Based Application: Toward a Tool Supported Stepwise Approach." *Proceedings of the 10th IEEE International Symposium on Web Site Evolution* (WSE 2008), Beijing, 2008, pp. 27–36. DOI: 10.1109/WSE.2008.4655392. 10

41. Martin, K. and Pamela, M. "Automated GUI Testing on the Android Platform." IMVS Fokus Report 2010, vol. 4, no. 1, pp. 33–36, 2010. 12

42. Mozilla.org. Mozilla Developer Network. "JavaScript."[Online]. https://developer.mozilla.org/en-US/docs/Web/JavaScript. 5

43. Nguyen, B.; Robbins, B.; Banerjee, I.; and Memon, A. "GUITAR: an innovative tool for automated testing of GUI-driven software." *Automated Software Engineering*, vol. 21, no. 1, pp. 65–105, March 2014. DOI: 10.1007/s10515-013-0128-9. 5

44. ourgrid.org. OurGrid. [Online]. http://www.ourgrid.org. 7

45. Pachal, P. (2014, July) Mashable.com. "Apple Pushes Hard Into Enterprise With IBM Partnership." [Online]. http://mashable.com/2014/07/15/apple-ibm/. 2

46. Parveen, T.; Tilley, S.; Daley, N.; and Morales, P. "Towards a distributed execution framework for JUnit test cases." *Proceedings of the 25th IEEE International Conference on Software Maintenance* (ICSM 2009), Edmonton, Canada, 2009, pp. 425–428. DOI: 10.1109/ICSM.2009.5306292. 3, 23

47. Penn, J. "Test iOS Apps with UI Automation: Bug Hunting Made Easy." *The Pragmatic Programmer*, 2013. 2, 12, 13, 28

48. Razak, A. and Fahrurazi, F. "Agile testing with Selenium." *Proceedings of the 5th Malaysian Conference in Software Engineering* (MySEC 2011). Johor Bahru, Malaysia, 2011, pp. 217–219. 27

49. Rogowsky, M. (2013, April) Forbes.com. "The Death of the PC Has Not Been Exaggerated."[Online]. http://www.forbes.com/sites/markrogowsky/2013/04/11/the-death-of-the-pc-has-not-been-exaggerated/. 1

50.   Rothermel, G. and Harrold, M.-J. "A safe, efficient regression test selection technique." *ACM Transactions on Software Engineering and Methodology* (TOSEM), vol. 6, no. 2, pp. 173–210, April 1997. DOI: 10.1145/248233.248262. 25

51.   Rothermel, G. and Harrold, M.-J. "Analyzing Regression Test Selection Techniques." *IEEE Transactions on Software Engineering*, vol. 22(8), 1996, pp. 529–551. DOI: 10.1109/32.536955. 3

52.   Square. (2011, July) Square. "iOS Integration Testing." [Online]. http://corner.squareup.com/2011/07/ios-integration-testing.html. 6

53.   Statista. (2014, October) Statista: The Statistics Portal. [Online]. http://www.statista.com. 1

54.   SWEBOK. (2014) IEEE Computer Society. [Online]. http://www.computer.org/portal/web/swebok/htmlformat. 9

55.   The Apache Software Foundation. (2013, April). Welcome to Apache Hadoop. "MapReduce Tutorial." [Online]. https://hadoop.apache.org/docs/r1.2.1/mapred_tutorial.html. 51

56.   The Apache Software Foundation. (2013, November) "Apache Hadoop." [Online]. http://hadoop.apache.org. 20

57.   ThoughtWorks. Frank. "Testing With Frank - Painless iOS and Mac Testing With Cucumber." [Online]. http://www.testingwithfrank.com. 6

58.   Tilley, S. and Floss, B. *Hard Problems in Software Testing: Solutions Using Testing as a Service (TaaS)*. Morgan & Claypool, 2014. DOI: 10.2200/S00587ED1V01Y201407SWE002. xv

59.   Tilley, S. and Parveen, T. (Editors). *Software Testing in the Cloud: Perspectives on an Emerging Discipline*. IGI Global, 2012. DOI: 10.1007/978-3-642-32122-1. xv

60.   Tilley, S. and Parveen, T. *Software Testing in the Cloud: Migration and Execution*. Springer, 2012. xv, 3, 23, 25, 28

61.   Tilley, S. Hard Problems in Software Testing. [Online]. http://www.hpst.net. xv

62.   Tilley, S. Software Testing in the Cloud. [Online]. http://www.stitc.org. xv

63.   Vollmer, A. "Tuneup JS." [Online]. http://www.tuneupjs.org. 14

64.   Weirich, J. "Rake -- Ruby Make." [Online]. http://rake.rubyforge.org. 17

65.   White, T. *Hadoop: The Definitive Guide* (3rd ed). O'Reilly Media / Yahoo Press, 2012. 20, 52

66.   Wong, E.; Horgan, J.; London, S.; and Agrawal, H. "A Study of Effective Regression Testing in Practice." *Proceedings of the 8th IEEE International Symposium on Software Reliability Engineering* (ISSRE), Albuquerque, NM, 1997, pp. 264–274. DOI: 10.1109/ISSRE.1997.630875. 3

67.   Yahoo! (2010, August) Yahoo! Developer Network. "Apache Hadoop: Best Practices and Anti-Patterns." [Online]. http://developer.yahoo.com/blogs/hadoop/apache-hadoop-best-practices-anti-patterns-465.html. 51

About the Authors

Scott Tilley is a Professor in the Department of Education and Interdisciplinary Studies at the Florida Institute of Technology, where he is Director of Computing Education. He is Chair of the Steering Committee for the IEEE Web Systems Evolution (WSE) series of events and a Past Chair of the ACM's Special Interest Group on Design of Communication (SIGDOC). He is an ACM Distinguished Lecturer. His main fields of interest are software engineering and computing education and his software engineering research lays at the intersection of software testing, cloud computing, and system migration. Scott's work in computing education focuses on computing literacy, educational technology, and STEM outreach. He writes the weekly "Technology Today" column for the *Florida Today* newspaper (Gannett).

Krissada Dechokul is a web and software developer from Thailand who has a passion for Apple technology, especially the iOS. He previously worked as a GIS programmer at ESRI (Thailand) Co., Ltd. He earned his Bachelor degree in Information and Communication Technology from Mahidol University and holds an MSE degree from the Florida Institute of Technology.

Printed in the United States
by Baker & Taylor Publisher Services